Deep Ocean

Tony Rice

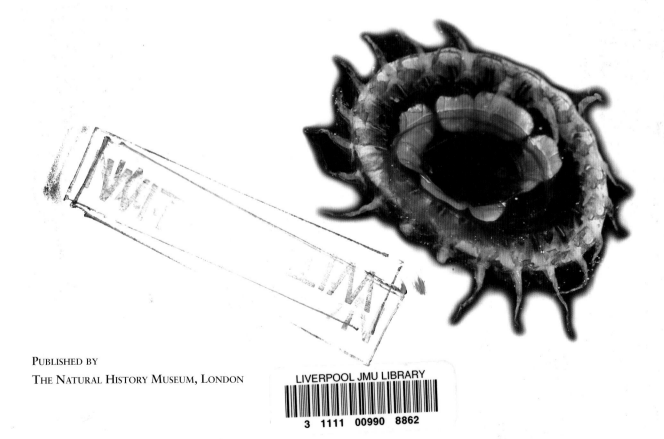

PUBLISHED BY
THE NATURAL HISTORY MUSEUM, LONDON

First published by The Natural History Museum,
Cromwell Road, London SW7 5BD
© The Natural History Museum, London, 2000
ISBN 0-565-09150-6

A catalogue record for this book is available from
the British Library

Edited by Jacqui Morris
Designed by Mercer Design
Reproduction and printing by Craft Print, Singapore

DISTRIBUTION

North America, South America,
Central America and the Caribbean
Smithsonian Institution Press
470 L'Enfant Plaza
Washington D.C. 20560-0950
USA

Australia and New Zealand
CSIRO Publishing
PO Box 1139
Collingwood, Victoria 3066
Australia

UK and rest of the world
Plymbridge Distributors Ltd.
Plymbridge House, Estover Road
Plymouth, Devon PL6 7PY
UK

Contents

Preface

The Earth is dominated by the oceans. They cover two-thirds of its surface to an average depth of about 4 km (2.5 miles) and to an extreme depth of more than 11 km (6.8 miles). They control the weather systems and the make-up of our atmosphere. Without them, no life would be possible on the planet – including our own. Yet for most of us the deep oceans and the strange animals that inhabit them are as alien and remote as the surface of the moon. Only 150 years ago the best informed scientists thought the depths of the seas were totally lifeless. We now know that animal life is to be found everywhere in the oceans, even at the bottom of the deepest trenches and around superheated water gushing through cracks in the sea floor. Indeed, many scientists believe that the deep seas may harbour more species than all of the Earth's other environments put together. This book traces the story of how this remarkable change came about and summarizes what we know, or think we know, of life in the deep oceans today.

The author

Tony Rice did his first degree and PhD in marine biology at the University of Liverpool. Apart from a 2-year fellowship at the University of Miami, he spent his working life in the UK, employed successively by Unilever, The Natural History Museum in London and the Natural Environment Research Council's Institute of Oceanographic Sciences, where he led the deep-sea benthic biology team for 26 years. He has published about 200 research papers, more general articles and books. Since his retirement in 1998 he has concentrated on writing books and acting as a marine environmental consultant.

Introducing the ocean

The Earth's oceans are vast, covering more than two-thirds of its total area and with an average depth of almost 4 km (over 2 miles). Life on Earth almost certainly first evolved in the ocean, so if the seas had not existed in the past, we would not be here now. And if they did not still exist, we would not have much of a future because currents in the modern oceans control the Earth's climate, while the ability of sea water to absorb greenhouse gases will largely control the amount of predicted global warming in the next few hundred years.

At one time, the depths of the ocean were thought to be totally lifeless. But scientists have known for well over 100 years that the whole of the ocean environment, down to the very greatest depths, at more than 11 km (7 miles), is populated by living organisms. The oceans provide about 170 times as much living space as all of the Earth's other environments – soil, air and fresh water – put together. The floor of the deep sea may harbour many times more species than these other environments. This idea is a recent one – only a few years ago the deep sea was thought to be populated by a rather small number of different species.

ABOVE **The Earth from space. The dominance of the oceans, and why the Earth is called the 'Blue Planet', becomes obvious from this viewpoint.**

The shape of ocean basins

All of the Earth's major land masses are surrounded by relatively shallow seas with fairly flat bottoms. These areas are called continental shelves and extend, on average, to about 60 km (37 miles) from the shore line. They represent only about one-twentieth of the total area of the oceans, but they are by far the richest parts biologically and provide us with most of the marine fish we eat.

At the outer edges of the continental shelves the sea bed falls away into the deep sea proper as continental slopes. These have gradients of about 1:40, which on land would be a fairly easy hill to cycle up. At a depth of about 3000 m (9800 ft) the gradient decreases to 1:100 or less on the continental rise. Together, the continental slopes and rises represent around one-quarter of the total surface area of the oceans. Then, at around 3500–4000 m (11,500–13,000 ft) depth the bottom flattens out even further to an almost imperceptible slope of 1:1000, or even less. This is the largest single environment on Earth, the abyssal plain, with depths down to

about 6000 m (19,700 ft) and underlying well over half of the ocean surface. Finally, in some rather small areas of the oceans, particularly in the Western Pacific, the sea floor drops away again, this time into elongated gashes, the trenches, with water depths of 10–11 km (6–7 miles).

The ocean basins are not simple depressions that increase in depth regularly from the margins to the middle. For one thing, the continental slopes vary enormously in steepness, in some places being dissected by dramatic canyons with near-vertical walls. They would be one of the most impressive sights on Earth – if only we could see them. If all the water could be drained away and we could fly over the dry ocean bed we would see an even more impressive feature. Interrupting the abyssal plain is a vast interconnected mountain chain, the mid-oceanic ridge system, extending over some 45,000 km (28,000 miles) and crossing all the major oceans except the North Pacific. The ridges rise 2000–4000 m (6600–13,000 ft) or more from the deep-sea floor, in places actually

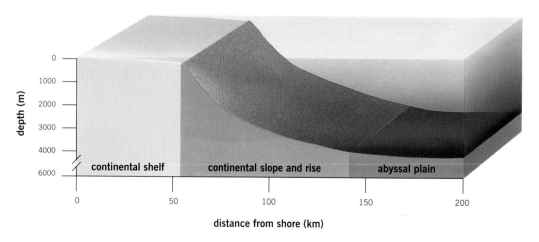

LEFT The main depth zones in the ocean. Note that the vertical scale is greatly exaggerated in this type of diagram. If you drew a profile of an ocean floor, e.g. the North Atlantic from New York to Lisbon, across this page to the same vertical and horizontal scales it would look like a straight line, because all the vertical inequalities would fall within a band less than one tenth of a millimetre wide.

depth (m)

0
1000
2000
3000
4000
6000

continental shelf   continental slope and rise   abyssal plain

0   50   100   150   200

distance from shore (km)

rising above the sea surface as oceanic islands. The ridges are the site of production of new sea floor, which rises as molten rock from deep within the Earth and then cools and moves away from the ridges at a few centimetres (an inch) a year like a great conveyor belt. The moving sea floor dips down into the deeper layers once more at the continental margins and also at the bottom of the trenches.

On its slow journey away from the ridges, the sea floor becomes carpeted with muddy sediment made up of volcanic ash and land-based material carried into the sea by rivers and winds, but mainly by the remains of billions of tiny animals and plants, which lived in the overlying waters and whose skeletons sank to the bottom after they died.

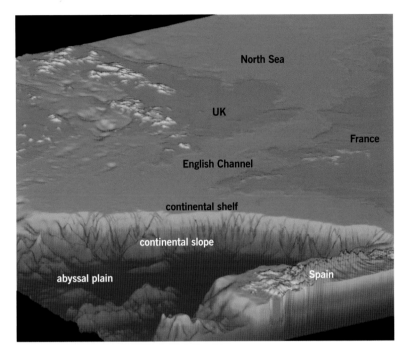

ABOVE **Computer-generated image of the continental margin off northwestern Europe as if viewed from above the Iberian Abyssal Plain and with the sea removed. The cliffs and canyons typical of the continental slope in this area produce one of the most dramatic geological features on Earth. (Vertical scale for subsea features is exaggerated.)**

RIGHT **The mid-ocean ridge system, stretching 45,000 km (about 27,960 miles) around the oceans and giving rise to new sea floor. The arrows show the directions of sea-floor spreading from the ridges.**

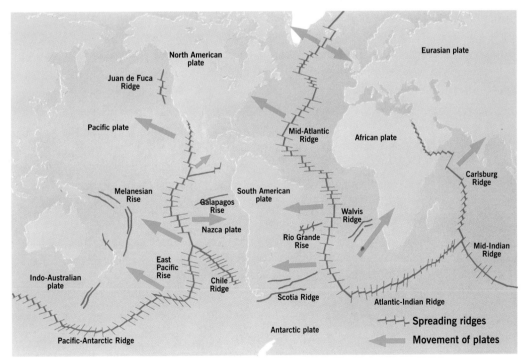

The problem of pressure

The weight of air in the Earth's atmosphere squeezes everything in it – animals, plants, rocks and so on – equally in all directions. This squeezing force, the atmospheric pressure, is at its maximum at sea level, where it is about 1 kg/cm^2 (15 lb/in^2). This is called conveniently, 1 atmosphere. At higher elevations, on mountains or in planes or balloons, the pressure is less because there is less weight of air above.

The weight of water in the sea also creates pressure. However, because water is so much heavier than air, the pressure changes much more rapidly with depth in the sea than it does with height in the atmosphere. In fact, in the sea it increases by 1 atmosphere with every 10 m (33 ft) of depth. So at a depth of 1000 m (3300 ft) the pressure is 100 times that at the surface, and at 10,000 m (33,000 ft) one thousand times that at the surface, about 1 tonne/cm^2 (6.6 tons/in^2). Many people find it difficult to imagine how any deep-sea animal can possibly withstand such a crushing pressure. But a fish or snail living in the abyss is no more conscious of the pressure of its surroundings than we are of the several tonnes that each of us is apparently carrying around as a result of the weight of the atmosphere on our bodies. In both cases, the internal and external pressures are the same and we become conscious of pressure only when we move to a different pressure quickly, for instance when we have difficulty clearing our ears in an aircraft or diving just a few metres beneath the surface with, or without, an aqualung.

The bodies of most marine animals are made up mainly of water, and because liquids are almost incompressible the animals suffer very little from the effects of quite large pressure changes. Many fish, including deep-sea ones, have gas-filled bladders, which allow them to stay more or less neutrally buoyant. The volume of such a gas-filled space is very sensitive to pressure change so that if the pressure is doubled the volume is halved, whereas if the pressure is halved the volume is doubled.

Fish with gas bladders living close to the surface have to be able to add or remove gas from their bladders as they move up and down, otherwise the volume of the bladder would change and upset the buoyancy.

The volume of the gas bladder of a herring, for example, moving from a depth of 30 m (98 ft) to the surface would increase fourfold. To experience a similar four-fold decrease in pressure and therefore a four-fold increase in the volume of its gas bladder, a fish living at a depth of 10,000 m (33,000 ft) would have to swim up to a depth of 2500 m (8200 ft). If, as in the example of the herring, it swam upwards for only 30 m (98 ft), that is to 9970 m (32,700 ft), the pressure would change only from 1000 to 997 atmospheres and the volume of its bladder would similarly change by only three one-thousandths, almost nothing. So deep-sea animals, including fish, can wander quite freely over large depth ranges without being concerned by the pressure changes involved.

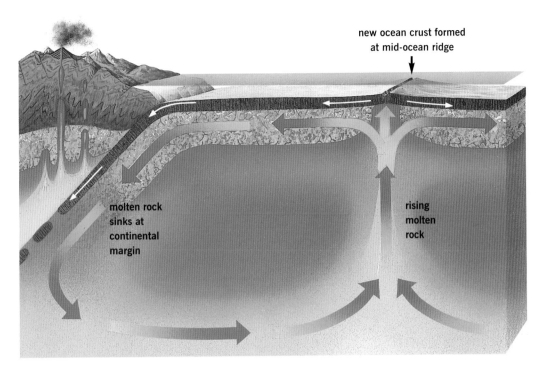

new ocean crust formed
at mid-ocean ridge

LEFT **Cross-section of
ocean floor showing
molten rock rising
beneath the mid-ocean
ridge and spreading
away either side.**

molten rock
sinks at
continental
margin

rising
molten
rock

The resulting sediment has accumulated on
the sea bed over millions of years, so most of
the irregularities are smothered by a layer of
mud, hundreds or even thousands of metres
deep. The upper layers of this sediment
provide the living space for the animals of
the deep-sea floor.

Sea water

The most characteristic feature of sea water is
that it is salty. This is because it is a complex
solution of many different chemicals. Except
near the mouths of rivers, where the salt
content may be very low, these chemicals
together make up between 33 and 37 parts
per thousand by weight of sea water
throughout the oceans, about 35 g (1 oz) of
salt in every litre (1.76 pt). Almost nine-tenths
of this salt is exactly that, common salt or

sodium chloride, but the remainder probably
includes all the chemicals that occur naturally
on Earth – and a few that do not occur
naturally, but which have been introduced to
the oceans by humans. Many of the chemicals
in sea water occur in very low concentrations,
or vary a good deal from place to place or
from time to time, as a result of the biological
processes going on within the oceans. But
several of the major constituents, including
sodium and chlorine, maintain very similar
concentrations throughout the seas, and have
done so for many millions of years. This
constancy means that, compared with fresh
water or dry land, sea water is a relatively
easy environment for animals and plants to
cope with physiologically. It is also why the
very first life on Earth almost certainly
evolved in the ancient seas.

Restless waters

If sea water is fairly constant in its chemical composition, it certainly is not as far as movement is concerned. The most obvious water movements to us are the waves and tides. But, although tidal currents can be detected even at the bottom of the deepest oceans, their main effects, and certainly those of waves, are restricted to very shallow layers. In the depths of the ocean the more important water movements are those of the global system of oceanic currents, which are so crucial in controlling the Earth's climate.

Surface currents

The surface currents are driven mainly by the winds. The result is a complicated system dominated by a series of gyres, or large circular currents, moving clockwise in the northern parts of each ocean and anticlockwise in the south. In the Southern Ocean, where there are no land masses to interrupt the currents, an eastward-flowing current, the West Wind Drift, constantly circles Antarctica. The best known of the gyres, in the North Atlantic, includes the Gulf Stream, which transports warm water from the Caribbean along the eastern seaboard of North America and then across the ocean to Europe. This gyre is completed by the Canaries Current in the Eastern Atlantic, which transports relatively cold water south and west, and therefore back towards the Caribbean. The Gulf Stream is not a simple current but a vast moving mass of water, which gives rise to complex swirls and eddies on either side. The enormous amount of heat transported by it helps to make the climate in North-western Europe much warmer than that at similar latitudes on the other side of the ocean. One of the major worries about global warming is that this system could be stopped. If this happened, Europe could paradoxically become much colder than it is today.

A similar gyre occurs in the Pacific, but the better known one here is the anticlockwise gyre in the southern part of the ocean because it includes the infamous El Niño phenomenon. The eastern part of this gyre consists of the Peru Current, which normally flows northwards along the coast of South America and then westwards across the Pacific towards Australia. The Peru Current is associated with rich plant and animal growth in the surface waters and very important fisheries. But every few years the system fails and the 'rich' Peru Current is displaced by a very 'poor' southward-flowing current, El Niño. The major changes in water flow associated with El Niño not only cause dramatic

BELOW **The main surface currents of the world's oceans. These are mostly driven by the winds, modified by the rotation of the Earth and the shapes of the land masses.**

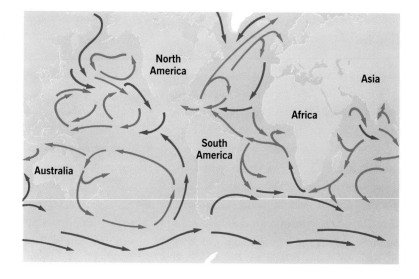

changes in the marine conditions off the coast of South America but may have important effects on the weather in many other parts of the world.

Deep circulation

Deep beneath the surface another current system gently stirs the waters of the ocean and this one is much more important for the animals of the deep sea. This is called the thermo-haline (heat and salt) system. These currents are driven by the sinking and rising of waters of different heaviness or density caused by differences in salt content and particularly temperature, rather than by the wind. Basically, very cold and salty (and therefore heavy) water sinks at high latitudes, especially in the North Atlantic and Pacific Oceans and around Antarctica, to be replaced by warmer and less salty water from lower latitudes. The result is a complex 'layer cake' of interleaved water masses flowing slowly under and over one another in the depths of the ocean. This deep circulation is important, not only because it mixes the water and keeps its chemistry more or less uniform, but also because it carries oxygen from the atmosphere into the deeper layers and makes life there possible.

Temperature

The surface temperatures of the oceans range from 40°C (104°F) or so in shallow tropical lagoons to — 1.9°C (27°F), the typical freezing point for sea water, in polar regions. If we could take a giant whisk and mix all the ocean waters together, the resulting average temperature would be no more than 3–4°C

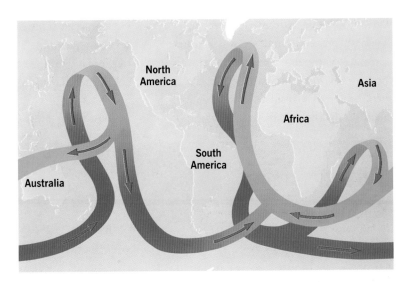

ABOVE **The deep ocean conveyor belt. The deep circulation is very complex, but this simplified version shows the main movements of warm and cold water.**

(37–39°F). This is because most of the water in the deeper layers is very cold. With the exception of the Mediterranean and Red Seas, which, because they are almost cut off from the rest of the oceans have rather warm deeper layers (around 13°C or 55°F in the Mediterranean), any warm water in the open ocean is restricted to a shallow, near-surface band. No matter how warm the surface layers are, between about 300 and 1000 m (980–3280 ft) beneath the surface the temperature falls to about 5°C (41°F) and then continues to fall more slowly with increasing depth. As a result, even beneath the hottest tropical regions the water at a depth of 2000–3000 m (6600–9800 ft) almost never rises above about 4°C (39°F), and most deeper waters are between 0 and 3°C (32–37°F) – with just one dramatic exception. Since the 1970s ocean scientists have become aware that extremely hot sea water gushes out of fissures in the underlying rocks in some places along the oceanic ridge systems. The water from these hydrothermal vents was originally ordinary 'cold' sea water, which percolated through the sedimentary layers to the hot rocks deep beneath the ridges and then rose up towards the surface through cracks in the rocks, rather like a water volcano. The water emerges at temperatures up to an incredible 300–400°C (570–750°F) but, because of the vast mass of surrounding cold water, the temperature drops to the normal 3–4°C (37–39°F) within a metre (3 ft) or so of the vent opening.

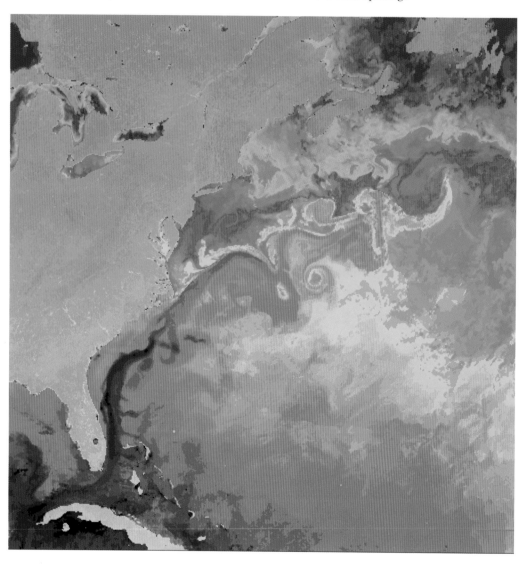

LEFT **Satellite image of the western Atlantic. Colours indicate different water temperatures and show the Gulf Stream along the eastern coast of the USA, and its swirls and eddies off New England.**

Development of oceanography

Early exploration of deep sea

Sailors and fishermen have known for centuries that the surface layers of the oceans, often hundreds of kilometres from the nearest land, can be teeming with life. They saw whales and fish at the surface as well as birds feeding there. They also knew that they could catch fish and other animals on lines and in nets well below the surface. But until about 150 years ago the deeper parts of the ocean were a total mystery. Scientists knew that the temperature of the water generally decreased rapidly with depth and that sunlight did not penetrate very far. They also knew that the pressure deep in the ocean would be enormous. They reasoned that life would be quite impossible in such a totally dark, icy-cold and bone-crushing environment, and until the middle of the 19th century most people believed that beyond a depth of a few hundred metres the oceans would be totally lifeless.

A few marine scientists harboured doubts that the deep ocean was lifeless, particularly when, in the 1850s and 1860s, broken submarine telegraph cables with animals firmly attached to them were brought back to the surface from great depths. The deep ocean was surely worthy of study, but it would be technically difficult and very expensive, the equivalent of space research in the second half of the 20th century.

ABOVE **Edward Forbes (1815-1854), a Manxman, was an excellent scientist, but thought the deep oceans were totally lifeless. He would have been delighted to learn that he was wrong.**

LEFT **HMS *Challenger*, a rather insignificant British naval vessel, carried out the world's first global oceanographic expedition from 1872 to 1876.**

H.M.S. Challenger.

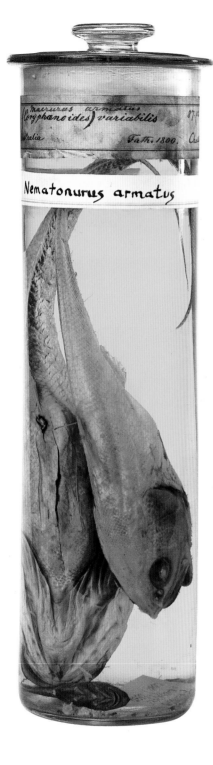

HMS *Challenger*

The first world-wide study of oceanography, the science of the deep ocean, was carried out from HMS *Challenger*, a British naval vessel with both sails and a steam engine, which, between 1872 to 1876, voyaged around the world with a staff of scientists of many different nationalities. The *Challenger* scientists were provided with the best equipment of the day, nets to be pulled through the water, trawls and dredges to be dragged across the sea bed, submarine thermometers, instruments to collect water and sediment samples – and enormous lengths of rope to lower this equipment into the sea. For three-and-a-half years the ship sailed through all the major oceans and covered 69,000 nautical miles (128,000 km or 79,000 miles). The scientists brought back vast amounts of information and tens of thousands of biological specimens. Some specimens came from as deep as 5700 m (18,700 ft) in the Pacific and demonstrated once and for all that animal life was to be found even in the very deepest parts of the ocean. Many of the animals collected had never been seen before. They were described in the official reports of the expedition, which filled 50 large volumes and 30,000 pages, the last volume being published in 1895, almost 20 years after the end of the voyage. The scientists who worked on the *Challenger* material were of many nationalities, giving oceanography an international flavour which continues today. And the *Challenger* samples, preserved in The Natural History Museum in London, are still studied by scientists from all over the world.

LEFT **The *Challenger* scientists collected tens of thousands of specimens which are still carefully conserved in The Natural History Museum, London, many of them like this rat-tail from the deep Pacific.**

BELOW **Dredge used in the *Challenger*, with teased out lengths of hemp rope to entangle the tiniest animals.**

ABOVE **Prince Albert 1 of Monaco (1848-1922)** a keen oceanographer, conducted many scientific cruises in his own yachts. *Princesse Alice II,* was built in 1898 for oceanography and had the latest technology on board.

BELOW **A modern British oceanographic vessel, RRS** *Discovery.* **Although she was originally built in 1962,** *Discovery* **was given a new lease of life when she was completely refitted in 1991/92.**

Modern oceanography

Following Britain's lead with the *Challenger* voyage, many other countries sent out major oceanographic expeditions, including France, Germany, Monaco, Russia, Sweden and the USA. At first these expeditions were more or less independent, although they often carried foreigners aboard, as the *Challenger* had done. With the passing years oceanography has become more and more international and most major research programmes now involve scientists, and often ships, from many different nations.

Apart from the use of wire cables instead of rope to suspend the equipment in the oceans, for many years the techniques used stayed much the same as those used on the *Challenger*. Although some of the fishing gear used today, such as trawls and dredges, is very similar to those used a century ago, in the last fifty years or so many revolutionary techniques have been used to investigate the oceans.

Trawling the bottom of the deep sea

Some operations in deep-sea science are much the same now as they were at the time of the *Challenger*. Trawling the sea floor is one of them. The *Challenger* scientists used a beam trawl, so-called because the mouth was kept open by a long wooden beam with skids at either end so that it would slide over the sea bed. Beam trawls are still used by commercial fishermen today and some deep-sea biologists also use them. But more often they use otter trawls, not invented until after the *Challenger* expedition. In these trawls the mouth is kept open by two steel plates, called otter boards, attached to either side of the net mouth. The boards are mounted at an angle to the direction of tow so that as they are pulled through the water they 'fly' outwards, a bit like kites, and stretch the mouth open. Vast lengths of thick hemp rope were used to pull the *Challenger* trawls, hauled in with noisy steam-driven winches. Today, thinner and stronger wire cables are used with much quieter electric or hydraulic winches. And whereas the *Challenger* scientists had no idea whether their trawls were working or not until they came back to the surface, modern oceanographers use electronics

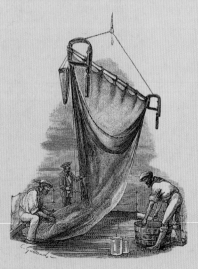

LEFT **The *Challenger* was supplied with vast amounts of equipment, including dredges and trawls to collect animals from the sea bed and more than 100 miles of rope. Here, a beam trawl is suspended over the ship's deck while two sailors deal with the catch.**

RIGHT **Otter trawl coming to the stern of a modern oceanographic vessel.**

and acoustics to tell them exactly what the gear is doing. But it still takes a long time. To trawl at 5000 m (16,000 ft) depth you need to pay out and haul in more than 10 km (6 miles) of wire, the whole operation taking well over 12 hours.

The anticipation when the net comes to the surface is just as nerve racking – even when the net is hanging over the deck with a bulging 'cod end' indicating a good catch, it could be filled with a load of pretty uninteresting rocks. The long wait is well worthwhile when the catch is as good as the one below, with lots of deep-sea sharks, rat-tail fish and other fascinating creatures from several kilometres beneath the ship.

LEFT **Otter trawl almost onboard, with the otter boards now clear of the water.**

BELOW LEFT **'Cod-end' of the trawl, where the catch accumulates, about to be emptied on to the deck.**

BELOW **A good catch. Now it has to be sorted, carefully documented and preserved.**

Sound

Sound travels very efficiently and rapidly through water, at about 1500 m/s (4900 ft/s) compared with 300 m/s (984 ft/s) in air. So sound, or acoustics, is used a great deal by ocean scientists. It was first used to measure the depth instead of lowering a line with a heavy weight on the end. In this technique, called echo-sounding, a series of sharp 'pings' is sent out from the ship and the depth is calculated from the time it takes for the ping to reach the bottom and the echo to come back to the ship. This calculation is done automatically so that, as the ship sails along, a profile of the bottom beneath the track appears on a screen or as a printed record. By sending out sound signals on either side of the ship's track it is even possible to produce a chart of the shape of the sea bed along a wide band beneath the keel. It is also possible to produce sounds that penetrate the sea-bed sediments, to be reflected by different layers within it. In this way, geologists obtain information on the structure of the sea floor deep beneath the bottom.

Sound can also be used by oceanographers to communicate with their instruments deep in the ocean. By building into instruments electronic 'ears' programmed to listen to particular sounds or patterns of sounds, the instruments can be made to respond to sounds sent to them from a ship thousands of metres away. One of the most frequently used techniques is the acoustic release. This uses a set of jaws that can be made to open in response to a sound signal. Acoustic releases are used with all manner of instrument packages, which are deployed on the deep-sea floor and left to collect samples or data for long periods, sometimes for a year or more. These instruments, for measuring water currents, collecting material sinking through the water column, recording temperature changes and so on, are usually attached to long wires, which are anchored to the sea bed with heavy weights, are kept vertical by large air-filled glass floats, and have an acoustic release somewhere in between. When the jaws are activated by the sound signal the wire is released from the weights and the floats bring the instruments back to the surface where they are picked up by the research vessel and the collected samples or data are analysed.

Submersibles and satellites

In the last 50 years oceanography has become dominated by electronics and computers. By lowering electronic sensors into the sea on the end of conducting cables,

LEFT **Deep-sea camera system being launched from a research vessel. The acoustic release, which will disconnect the weights, is in the centre of the triangular frame.**

vast amounts of information about the deep ocean, such as the water temperature and its chemical composition, can be sent back to the ship's computer without ever bringing a sample on board. Conventional cameras and video cameras lowered into the sea in pressure-resistant cases are widely used to obtain information about how deep-sea animals live.

Most oceanography is still carried out by lowering equipment into the ocean from a ship, but submersibles, really mini-submarines, are becoming increasingly used. Some, like the ones used initially to investigate the wreck of the *Titanic*, are unmanned robots or ROVs (remotely operated vehicles) carrying cameras and other equipment to the sea floor. They are tethered to the ship by a cable through which their operations are controlled. There are also many manned submersibles in which scientists can travel into the depths to make direct observations and even conduct complex experiments. *Mir*, meaning peace, is such a submersible, and can dive to the bottom of the abyss at depths of 6000 m (19,700 ft). Submersibles are now being developed that will be able to roam the oceans unattended. They will oscillate between the surface and the deeper layers, relaying data back to land-based laboratories via satellites, and finally return to their home port. And just as space satellites orbiting the Earth can send information back about our atmospheric weather systems, they can also gather data about the oceans much faster than from a ship. In this way oceanographers can produce detailed maps showing the distribution of sea temperatures, the amount of plant growth in the water and even the height of the sea surface to within a few centimetres.

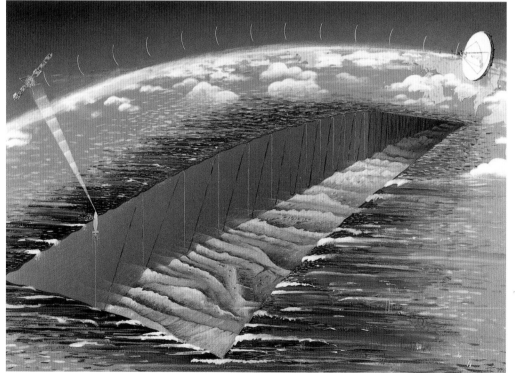

ABOVE **AUTOSUB, an underwater vehicle capable of wandering the oceans alone and sending information back to land automatically.**

LEFT **Artist's impression of AUTOSUB at work. AUTOSUB will oscillate up and down in the open ocean, sending its data via satellites back to shore-based laboratories each time it comes to the surface.**

Vertical life zones in the ocean

As oceanographers started to study life in the deep ocean they soon discovered two important features. First, although animals were to be found at all depths in the seas, there was a clear decrease in the amount of life with increasing depth, with a particularly marked fall between about 100 and 1000 m (330 and 3300 ft). Second, although some species could be found over very wide depth ranges, most of them were much more restricted in their vertical distribution and lived in fairly distinct communities, which could be found at the same depths over very wide areas of the oceans. Gradually it became clear that these communities live in quite well-defined horizontal slices or zones, some very thin and others extremely thick. Although the boundaries between the zones are not always very obvious, and some animals regularly move from one zone to another, the broad zones can be recognized anywhere in the ocean.

There are three main mid-water zones. The shallowest, and thinnest, is called the epipelagic, simply meaning upon the sea. It extends from the surface down to about 100–150 m (330–500 ft). Its most important feature is that it is lit by sunlight, which allows plant growth, whereas the whole of the deeper layers are too dark for this. The epipelagic zone impinges on the sea floor only in shallow inshore waters on the continental shelves. In all the deeper parts of the ocean its lower margin simply merges with the second depth zone.

The second zone is the mesopelagic, meaning middle sea. It is also often called the twilight zone because, while its upper regions, where it meets the epipelagic zone, are still quite light during the day, its lower boundary

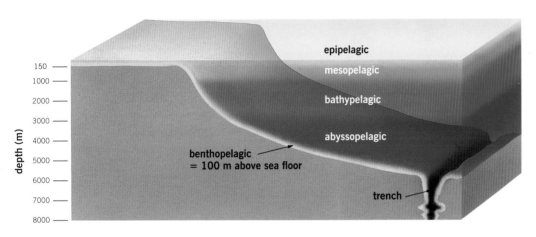

epipelagic

mesopelagic

bathypelagic

abyssopelagic

benthopelagic = 100 m above sea floor

trench

depth (m)

150
1000
2000
3000
4000
5000
6000
7000
8000

LEFT **The main vertical life zones in the ocean. Almost all animal life in the deep sea is dependent for its existence on phyto-plankton growth in the extremely thin sunlit surface layer.**

at about 1000 m (3300 ft) is perpetually dark because none of the Sun's rays can penetrate this far. Many of the animals in this strange twilight world produce their own light called bioluminescence (living light), which they use as camouflage or to find, or attract, potential food or mates. The mesopelagic zone impinges on the sea floor on the outer parts of the continental shelves and the upper part of the continental slopes. In the deep sea proper the mesopelagic zone overlies the vast, dark waters of the last great depth zone.

The third zone receives no sunlight at all, so it is always absolutely dark apart from the occasional flashes of ghostly blue light produced by bioluminescent animals. It is rather arbitrarily divided by ocean scientists into the bathypelagic (deep sea), which extends down to about 3000 m (9900 ft), the abyssopelagic (bottomless sea) from 3000 to 6000 m (9900–19,800 ft), and the hadal (unseen) for the ocean trenches. But we can lump them all together into the abyss.

Underlying the water column is the sea floor itself. This is called the benthic zone and the animal communities living there are called collectively the benthos, names that are derived from a Greek word meaning deep. The benthic zone is the thinnest and most clearly limited zone of all, because most benthic animals live either on the surface of the sea-bed sediments or within the top few centimetres of bottom mud. Even this zone is blurred because some sea-floor animals, particularly the shrimps and fish, move away from the bottom from time to time, sometimes swimming tens or even hundreds of metres above it.

The intensely cold, dark and bone-crushing deep-sea environment seems totally alien and unimaginably hostile to us. Yet for many creatures this is home, and they would find our world, or even the upper sunlit layers of the sea, just as alien. Without these upper sunlit layers, however, most deep-sea animals would be in big trouble, as we shall see.

Pastures of the sea

Plants of the surface layers

Plants in the sea, like their land-living counterparts, require the energy of sunlight for photosynthesis, the process of converting simple inorganic chemicals into complex organic substances, which animals need but cannot make. Light does not pass easily through water and is rapidly absorbed and scattered by particles in suspension. So even in the clearest oceanic waters, at a depth of about 100 m (330 ft) the visible sunlight is reduced to no more than about one-hundredth of its intensity at the surface. In the generally much murkier inshore waters the drop-off in light intensity is even more rapid. At the same time, the light becomes relatively less red and more blue with increasing depth. The result is that photosynthesis in the sea is restricted to the upper few tens of metres at most. Consequently, seaweeds attached to the bottom can exist only where the depth is less than this limit for photosynthesis, a very narrow strip close to the shore.

Phytoplankton

The vast majority of the plants in the oceans are tiny single-celled algae, which float in the sunlit surface layers and are collectively called the phytoplankton. There are about 4000 phytoplankton species, which is not many

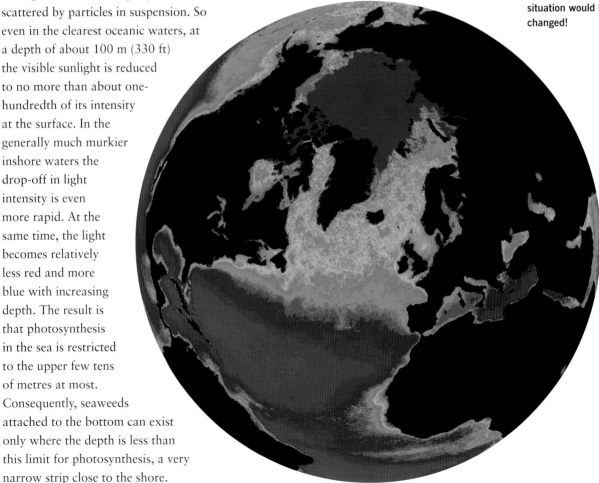

BELOW **Satellite image of the Atlantic showing high phytoplankon concentrations in the northern half. It would be impossible to obtain this sort of information from ships; it would take weeks to collect the data – by which time the situation would have changed!**

compared with the estimated 250,000 species of green plants on land. Two of the most abundant and important phytoplankton groups are the coccolithophorids and diatoms, whose dead remains form major parts of the deep-sea sediments in many areas. Coccolithophorids are really tiny, most of the 150 or so species being less than one-fiftieth of a millimetre (one-thousandth of an inch) across and therefore totally invisible to the naked eye. Their single cells are enclosed within a calcareous, or chalky, shell made up of a series of little circular plates called coccoliths. These are very characteristic features of many marine sediments, being a major constituent, for instance, of the famous white cliffs of Dover, England, which were formed under the seas of the Cretaceous period 142–165 million years ago. Although most coccolithophorid species live in the warmer seas, one species, *Emiliana huxleyi*, is found in all seas. Although individuals of this species are impossible to see without a microscope, its populations can be so vast, both in numbers and area, that they are detectable from satellites. Most diatoms are also tiny, but some of them can reach a diameter of a millimetre (0.04 in) or more and they come in a bewildering variety of beautiful forms. All are single-celled, but as they multiply by cell division the resulting daughter cells often stay attached to one another to form long chains, sometimes several millimetres long. Their external skeletons are made of glass-like silica rather than chalk and they often dominate sediments in the very deepest parts of the oceans, where chalky remains tend to dissolve.

Plant production in the oceans

The total plant production in the oceans each year is about 20,000 million tonnes (19,700 tons). It is not evenly distributed because, just like land plants, marine plants require not only adequate light and temperatures, but also a supply of nutrients, particularly nitrates and phosphates, in order to grow. So plant growth in the oceans is particularly rich over the shallow waters of the continental shelves where these chemicals are carried into the sea by rivers and where nutrient-rich bottom

BELOW
A coccolithophorid, *Coccolithus pelagicus* from the South Atlantic, showing the ball of tiny plates enclosing the protoplasm.

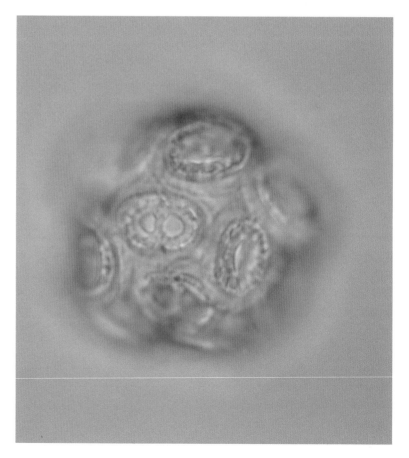

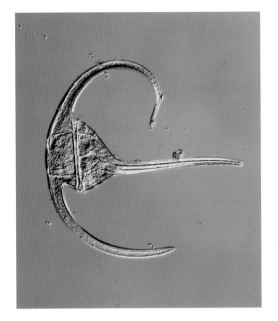

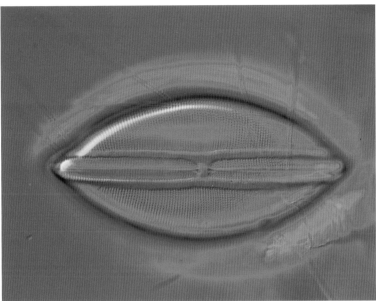

waters are stirred into the sunlit surface layers by currents and storms. These are also the areas where most of the world's fisheries are based, including those for cod and haddock, flatfish (such as plaice, flounder and sole), and mid-water fish (such as herring and mackerel).

In the surface waters of the deep ocean the nutrients tend to be in much shorter supply, so the central parts of all of the great oceans are many times less productive than the continental shelves. But there are a few places in the deep ocean where the phytoplankton growth is much higher than usual because, for complex reasons, nutrient-rich deep water is brought up to the surface. These upwelling zones are concentrated off the western coasts of Africa, and North and South America, but with tongues of high productivity stretching across the Western Pacific and all around the Southern Ocean. Not surprisingly, these are the sites of the great oceanic fisheries for wide-ranging fish such as tuna. In the past the Southern Ocean was also the scene of a huge whaling industry.

Phytoplankton production varies not only from place to place but also with the time of year. In tropical waters where the surface sea temperatures stay high all year round, and where there is plenty of sunshine almost every day, the seasonal effect is minimal. But in temperate and high latitudes, in both the Northern and Southern Hemispheres, the tiny plants can grow and reproduce only when the water temperature is high enough, when there are plenty of nutrients, and when there is plenty of sunshine. Just as on land, in these regions the plants lie more or less dormant over the winter, grow and reproduce rapidly in the spring, and continue to grow during the summer months until they either run short of nutrients or the water temperature and sunlight decrease in the autumn.

Diatoms ABOVE LEFT **and dinoflagellates** ABOVE RIGHT, **representatives of two of the main groups of planktonic plants in the oceans.**

Animals of the surface layers
Plankton

The seasonality in phytoplankton abundance naturally dominates the lives of the animals. Like the plants, the populations of relatively short-lived, tiny herbivores and the carnivores that feed directly on them tend to die down during the winter and then increase in abundance dramatically in the spring, once the plant populations have started to grow again. During the food-rich summer months many of them produce several successive fast-growing broods of young before finally producing, in the autumn, a much slower growing and longer-lived population, which will over-winter and eventually give rise to the first generation of the following year. Many of the much larger and long-lived animals, including fish, which breed only once a year, produce their young in the spring or early summer to coincide with the abundance of food at this time.

The surface layer of the sea, the epipelagic zone, is also the home of all the marine plant eaters or herbivores, the ocean's equivalent of cows, sheep, deer and so on. The planktonic plants are tiny, so the animals that eat them also have to be very small. Lots of different animal groups feed on the microscopic phytoplankton cells, including the young larval stages of many species, which, as adults, are far too big to feed on the tiny plants. By far the most important group of marine herbivores are the copepods, small crustacean relatives of the familiar crabs, lobsters and shrimps. Like all crustaceans, and like other arthropods such as the insects, spiders and so on, copepods have an external

ABOVE **A typical oceanic copepod.**

skeleton, which has to be shed or moulted from time to time to allow the animal to grow. These moulted skins, the copepod's faecal pellets containing incompletely digested phytoplankton, and, of course, the dead remains of any copepods that are not eaten in the surface layers, all contribute to the rain of small food particles, which sinks through the ocean to fuel the deeper layers.

All copepods are tiny, ranging in length from only about 0.5 mm (0.02 in) to no more than about 17 mm (0.7 in). They are all built on roughly the same body plan, with a cigar-shaped body from which spring a pair of long feelers, or antennae, at one end and a short, narrow tail at the other end. The body carries a series of appendages including several that are used to 'row' the animal through the water (copepod means oar-footed). As the copepods swim they collect tiny phytoplankton cells suspended in the water and pass them through their mouths and into their stomachs.

Predators

Copepods and the other small herbivores can be extremely abundant; even in the open ocean there may be hundreds or even thousands of them in every cubic metre of water near the surface. This is just as well, because the shallow waters also teem with predators, which make a living by eating the herbivore swarms. In general, in the mid-water world animals tend to eat food that is no more than about one-tenth as long as themselves. So the copepods are eaten mostly by predators ranging in length from a few millimetres (0.1 in) to a few centimetres (an inch or so). These include the voracious arrow worms, long, very slender and exclusively marine creatures, which, as their name suggests, can move extremely rapidly and grab their prey in fearsome, vice-like jaws. Less agile, but equally dangerous to tiny creatures such as the copepods are dozens of species of small jellyfish. Like many of their larger relatives, they have stinging cells, which shoot barbed and poisonous projectiles into their prey. There are many squid and fish that are small enough to feed on the little herbivores: some are the juvenile stages of species that will feed on much larger prey when they reach adulthood; others grow to only 4–5 cm (1.6–2 in) long at maturity and feed on tiny prey throughout their lives.

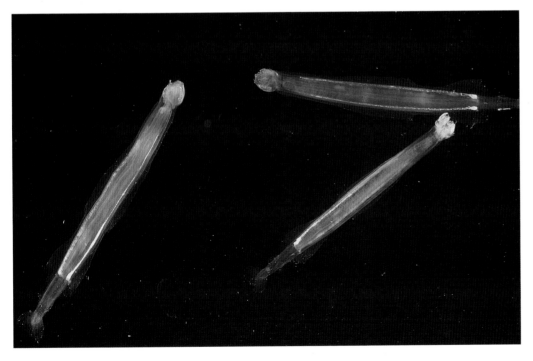

LEFT **Arrow worms or Chaetognaths, meaning 'bristle-jawed'. These little creatures, usually no more than 2 or 3 cm (1–1.5 in) long, are major predators on the other planktonic animals, particularly the copepods.**

RIGHT *Pantachogon* sp., one of the many hundreds of species of tiny predatory jellyfish in the oceanic plankton.

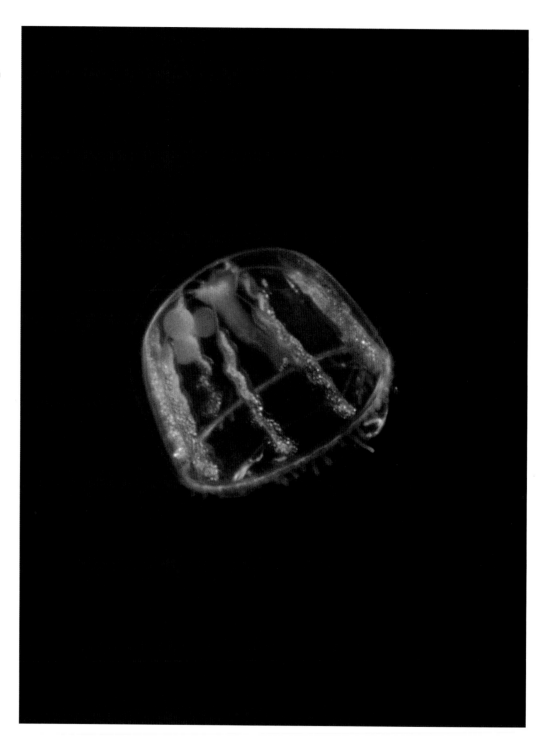

These tiny predators are themselves the food for the next size category – middle-sized fish, squid, shrimps, jellyfish and so on up to a few tens of centimetres long. These are, in turn, food for the really large hunters, mainly squid and fish up to several metres long, including tuna, marlin and some of the oceanic sharks. These are the top predators of the open ocean, the equivalent of lions and wolves on land. Like their terrestrial equivalents, they are relatively uncommon because, with so many links in the food chain, it takes a large area of phytoplankton production to support them.

Some large marine animals break the unwritten rule relating size of food to size of eater. At one extreme, some large predators, including some of the sharks, feed on animals that may be as big or bigger than themselves by biting mouth-sized lumps out of them. And in the deeper layers some fishes can even swallow creatures larger than themselves. At the other end of the scale, some very large animals feed on creatures that are many hundreds or even thousands of times smaller than themselves. Well-known examples are the huge plankton-feeding whale shark and basking shark, and, most famous of all, the blue whale. Weighing up to 100 tonnes (98 tons) or more, the blue whale maintains its vast bulk by swallowing enormous quantities of krill, little shrimps no more than 3–4 cm (1–1.6 in) long. Krill are herbivores, feeding directly on the plant plankton, so there are only two steps in the food chain from the tiny photosynthesizing phytoplanktonic algae to the largest animal that has ever existed on Earth.

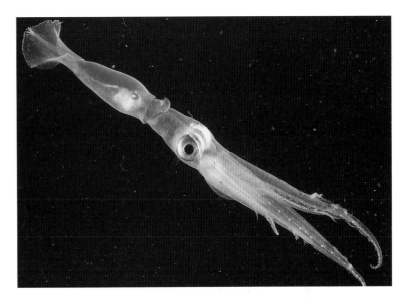

ABOVE **Krill,** *Euphausia superba,* **the main food of the blue whale. Krill belong to a group of small planktonic shrimp, the Euphausiacea, found throughout the oceans, of which there are only about 70 species worldwide.**

TOP *Chiroteuthis,* a slender, big-eyed deep-sea squid.

Mid-water world – from twilight zone to abyss

For the animals of the deeper layers of the water column where there is no plant growth at all, finding enough food and avoiding being eaten are among the main problems they face. Although the mid-water communities of the deep ocean are not very species-rich considering the vast size of this environment, they include representatives of almost every major animal group on Earth, from tiny single-celled organisms only a fraction of a millimetre (0.04 in) across, through all the invertebrate phyla, to fish several metres long. Most of them are in the middle size range, from about a centimetre (0.4 in) to several centimetres long, and are dominated by crustaceans, fish and a variety of animals with soft and rather jelly-like bodies. It is difficult for us to imagine what it must be like to live perpetually in the middle of the deep ocean, surrounded on all sides by seemingly endless water. By looking closely at a few of the animals that live in this strange environment we can get an idea of some of its advantages – and some of its disadvantages.

Shape and colour

Because they never come into contact with a physical barrier of any kind, such as the sea surface or the bottom, many mid-water animals have bizarre body shapes, more or

LEFT *Periphylla periphylla*, a deep-sea coronate jellyfish.

OPPOSITE **Deep sea jellyfish, *Atolla* sp. from the twilight zone.**

less spherical and often with long and delicate extensions. Many of them do not need a rigid skeleton because they spend all their time suspended in water. This is why soft and watery bodies seem to be so popular in the mid-water world. These jelly-bodied creatures include not only more or less conventional jellyfish and their relatives, the siphonophores, but also comb-jellies and salps, and even some members of groups such as the squid, which usually have rather firm and powerful bodies.

Whether you have a soft and delicate body or not, it is important not to be seen too easily, both to avoid being eaten by predators and so that you can get close enough to your own preferred food animals to catch them. So in the upper parts of the mesopelagic zone, where there is still quite a lot of sunlight, many animals have more or less transparent bodies so that an observer looks straight through them. As an alternative to transparency, many of the little fish of the upper mesopelagic, such as the hatchet fish and lantern fish, have silvery sides that act as mirrors and light organs along their bellies. At the darker and deeper levels, where bioluminescence is the only light around, these camouflage techniques do not work. Instead, the deep-living shrimps tend to be

BELOW *Acanthephyra* sp., a typical deep red shrimp from the lower mesopelagic and bathypelagic zones.

LEFT The fang-tooth or ogrefish, *Anoplogaster cornuta*, from the bathypelagic zone. Like so many other fish from the dark mid-water realm, *Anoplogaster's* body is dark-coloured so that it does not reflect the blue bioluminescence produced by potential predators.

dark red and the fish uniformly black or brown. These dark colours do not reflect the blue colour of most bioluminescence so they have the same effect as an organism being transparent in the sunlit layers.

Catching food

As the light decreases with increasing depth, so does the amount of potential food. Few deep mid-water animals can afford to be very fussy about what they eat and most of them feed on whatever they can catch. To do this, many of the tiny fish have mouths full of formidable-looking teeth, while one important deep-sea group, the anglerfish, also have 'fishing rods' with bioluminescent lures to entice potential food organisms to come close to their ferocious jaws. Many of these fish have large mouths so that they can swallow quite big prey. This is taken to extremes in the gulper eels, which, as their

ABOVE *Eurypharynx pelicanoides,* an appropriately-named gulper eel from the lower mesopelagic and bathypelagic zones. *Eurypharynx* grows to about 75 cm (30 in) long, but despite its large mouth it has poorly developed teeth and eats mainly relatively small shrimps and fishes. But other gulper eels have much bigger teeth, grow up to 2 m (6.5 ft) long and can swallow very large fishes.

name suggests, are highly evolved swallowing-machines. They have extremely elongated bodies, sometimes more than a metre 3ft long and most of the length consists of a very thin and tapering tail. The enormous and gaping mouth opens into a large stomach, enabling some of these eels to swallow animals even longer than themselves.

Day in and day out these curious animals, and hundreds more like them, live out their lives in this strange and vast world of constant cold and darkness, lit only by the ghostly light of luminescence and with no knowledge either of the world above them, with its dramatic changes from night to day and the passage of the seasons, or of the equally strange world of the sea floor hundreds or thousands of metres beneath them.

Vertical migration

Considering the relative abundance of food in the surface layers of the ocean, phytoplankton for the herbivores and lots of different animals for the carnivores, you might expect that any creature capable of living in this rich 'soup' would be very unwilling to leave it. But you would be wrong, because one of the most puzzling features of mid-water life in the oceans is that many animals do not stay at the same level all the time, but spend an enormous amount of time and effort moving up and down, often over hundreds of metres, as regularly as clockwork every single day of their lives.

Dawn to dusk

The existence of these amazing journeys, called collectively diel vertical migration, first came to light in the early days of marine biology when scientists noticed that plankton nets, fished near the surface at night, commonly collected more animals belonging to many more species than they did during the day. At first, some biologists thought that the animals could see the nets during the daytime and avoid being captured, whereas they could not do this at night. As collecting techniques improved, particularly with the development of nets that could be both opened and closed at any depth

beneath the surface, it became obvious that many species occupy quite different depth zones at night and during the day.

The details of the migration are different for different species, but the general pattern is for the animals to spend the daylight hours at depth and to move upwards, towards the surface, at around dusk. Some species then spend the night fairly concentrated at their chosen shallow depth before making the return journey down to their daytime level around dawn. Others seem to spread over a wide depth range during darkness with the deeper members of the population rejoining the shallower ones around dawn, before the whole population moves down once more to the day depth. These movements are clearly related to the changing light levels in the sea so they are restricted to the upper layers, which receive detectable light from the Sun. But deep-sea animals must be very sensitive indeed to light, for in the clearest waters of the open ocean some migrators spend the day at depths of 1000 m (3300 ft) or more where, even at mid-day, the light level is extremely low.

The range of the migrations, that is the difference between the daytime and night-time depths, tends to increase with the size of the animals involved. So the small planktonic animals less than about 1 mm (0.04 in) long migrate through only 10–20 m (33–66 ft). Many of the smaller herbivorous copepods undertake these sorts of migrations. The larger plankton, up to about 20 mm (0.8 in) in length and including some of the larger copepods, may move through 100–300 m (330–990 ft), while the even bigger and more powerful swimmers several centimetres (an

inch or so) long may regularly migrate up to 1000 m (3300 ft) or even more. Among these strong swimmers are many of the big mid-water shrimps such as *Systellaspis debilis*, which is 6 cm (2.4 in) long. By day *Systellaspis* lives in the mesopelagic zone at around 500–600 m (1600–2000 ft) deep but each night it swims up to within 50 m (160 ft) or so of the surface. By shrimp standards even this is fairly modest, because several relatives of *Systellaspis* regularly migrate through more than 1000 m (3300 ft).

Lantern fish

The champion migrators are certainly among the mid-water fish, particularly the lantern fish. Mostly small, spindle-shaped fish between about 5 and 10 cm (2–4 in) long, the lantern fish get their common name from the amazing array of light organs they carry along their flanks and bellies. There are about 250 species of lantern fish worldwide; all of them probably migrate and most of them through several hundreds of metres, but the record holder seems to be

ABOVE **These shrimp, *Systellaspis debilis*, spend the daylight hours in the mesopelagic zone but swim almost to the surface every night.**

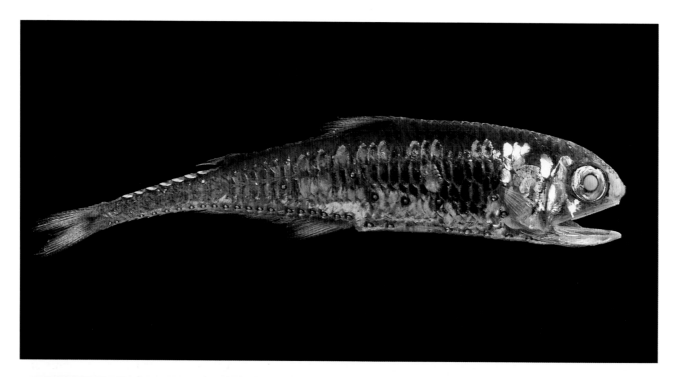

OPPOSITE TOP **Myctophum punctatum,** a mesopelagic lantern fish. Lantern fishes range from about 3 to 25 cm (1 to 10 in) in length and are found throughout the oceans. They get their common name from the array of light organs they carry.

OPPOSITE BOTTOM **Lepidophanes** sp., another lantern fish, showing the light organs along its ventral surface.

Ceratoscopelus warmingii, which lives at mid-latitudes in the deep North Atlantic. This diminutive little fish, only about 6 cm (2.4 in) long at most, lives at a depth of about 1600–1800 m (5300–5900 ft) during the day. Every single night it undertakes a 3-hour journey that will carry it up to within 100 m (330 ft) or so of the surface; and early every morning it takes another 3 hours to swim back down again. Why on earth do *Ceratoscopelus* and all the innumerable other vertical migrators undertake these surely exhausting journeys? After all, considering the size of the animals involved, many of these migrations are equivalent to a human being running a marathon every evening after supper – and another one every morning before breakfast. The short answer is that we do not know, although there have been lots of possible explanations suggested. One of the oldest ideas, and still one of the most popular, is that the herbivorous animals have to spend a lot of time close to the surface because that is where their food, the phytoplankton cells, are concentrated. But they must also avoid being eaten by the carnivores that feed on them. Where these predators are visual feeders, the prey animals would clearly be wise to avoid the sunlit surface waters during the day when they can be seen, and caught, very easily. This is exactly what they do, spending the daytime in the darker deeper layers and coming to the surface to feed only during darkness.

Deep migrators

This explanation does not account for the vertical migrations of the animals that live at much deeper levels. They live the whole of their lives in very low light levels, where visual feeding is probably not very important. So why do they migrate? Perhaps, like the shallow-living plant eaters, it is to maximize their chances of getting enough food, but at the same time to reduce their chances of being eaten. In general the number of animals in the sea decreases with increasing depth, so the nearer the surface an animal lives the more food will be available. However, potential predators will also be more common. Maybe, even for animals living at very deep levels, there is an advantage to be gained by dividing their time between relatively shallow depths, where food is more abundant but so are enemies, and safer deeper layers where food is scarcer. Whatever the explanation, every single day millions and millions of tonnes of living organisms all over the oceans move up and down over tens or hundreds of metres like an enormous biological piston – exactly as their predecessors have done for millions of years.

Bioluminescence

We have already seen that sunlight penetrates no deeper than about 1 km (0.6 miles) into the sea, even in the clearest oceanic waters. So why do many deep-sea animals, both in mid-water and on the bottom, have well-developed eyes? The answer is bioluminescence, which is produced by many organisms in the oceans. On land, just a few animals, such as fireflies and glow-worms, can produce light. In the sea, thousands of species, ranging from tiny bacteria and dinoflagellates to fish and squid, can do it.

Light-producers live at all depths, from the surface of the sea to the bottom of the ocean trenches. Even the strongest bioluminescence is no brighter than moonlight, so at shallow depths it is significant only at night or in very murky conditions. In contrast, in the deep ocean it is enormously important because it is the only light that most of the animals ever see.

Protection from predators

Why do so many marine animals, and particularly deep-living ones, produce light? In many cases we simply do not know, but here are a few well-informed guesses. In the upper few hundreds of metres of the water column, where sunlight is still important, many animals certainly use light as camouflage to protect them from predators. The hatchet fish are excellent examples. These little fish, usually no more than a few centimetres (an inch) long, are extremely flattened from side to side and are covered in highly reflective plates so that they are like vertical mirrors. Viewed from the side, they are more or less invisible. Despite being so thin, a potential predator looking up at them from below would be able to see them very easily silhouetted against the dim light from the surface, if it were not for the presence of a series of light organs shining down from the fish's belly. These organs produce a light that matches the intensity of the incoming natural light, but the really clever hatchet-fish trick is that the light they emit does not simply shine

RIGHT **Hatchet fish,** *Argyropelecus aculeatus,* **viewed from the side. These greatly flattened fish from the mesopelagic zone have silvered sides that act as mirrors, and light organs along their ventral surface to camouflage their silhouette against the downwelling light.**

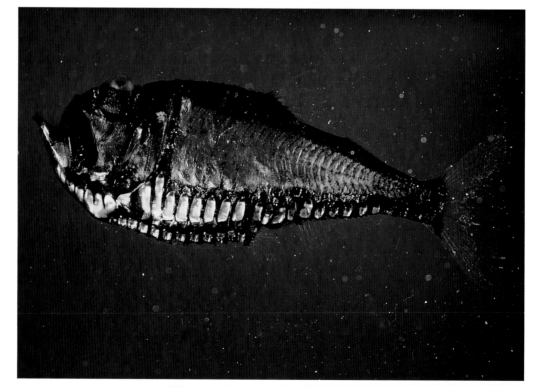

down vertically like a torch. Instead, within each light organ there is a complex system of mirrors and reflective filters so that the light released also matches exactly the angular distribution of natural light in the sea.

Many other animals in the twilight zone use mirrors and light organs to deceive predators, although few of them as effectively as the hatchet fish. Many more, even in the pitch-black regions below the limit of sunlight, use light to confuse or frighten predators rather than to hide from them. These animals tend to produce light suddenly when a predator threatens or grabs them, the apparent 'intention' being to shock the predator into letting go. An extension of this technique, used by some shrimps, worms, squid and even fish, is to release a luminous decoy to attract the predator's attention while the potential food animal escapes into the darkness. Some worms and brittle stars actually discard an unneeded light-emitting piece of their body to act as the decoy, the rest of the animal scurrying out of harm's way. Many others squirt out a smoke screen of luminous particles, which hangs like a bright cloud in the water to distract the predator for a few life-preserving seconds.

Finding mates

Far from making their owners invisible, or at least less conspicuous, the light organs of many species seem to have quite the opposite purpose, to advertise the animal's presence, particularly to a potential mate. Many small crustaceans, as well as some fish and squid, have patterns of light organs that are unique to their species and may even have quite different patterns in males and females so that mate-seekers do not waste their time with members of the same sex.

Some deep-sea animals seem to use light to attract food as well as mates. The angler fish are classic examples. There are many different angler-fish species, most no more than 5–6 cm (2–2.5 in) long, and in most species the males are very much smaller and less well-developed than the females, often spending their adult lives permanently attached to, and feeding from, a female, almost like a parasite. The females have light-emitting bacterial cultures in little sacs on the end of fishing-rod-like structures above their enormous tooth-filled mouths. It is easy to see how a potential food animal, attracted to the luminous lure, could be snapped up in the angler's jaws. But if the male anglers are also attracted to the lure, we still do not understand how they survive to become mates rather than food.

OPPOSITE *Gigantactis* sp., a ceratioid angler fish from the bathypelagic zone, with a biolumi-nescent lure at the end of the 'fishing rod'.

BELOW *Watasenia scintillans*, a deep-sea mid-water squid which, as its name suggests, scintillates with blue light organs.

How is bioluminescence produced?

Bioluminescence is not very different from the light from a candle or from burning paper or wood. In these cases a chemical reaction takes place in which the substrate (wax, paper or wood) is oxidized (burnt) to produce various other substances: gases such as carbon dioxide, and solids in the smoke and ash. The process also involves the release of a lot of energy, much of it as heat and some as light. In a similar way, bioluminescence involves oxidation of a substrate, luciferin, helped, or catalysed, by another chemical, luciferase. In this case, almost all the released energy comes out as light rather than heat, so you cannot get your fingers burned by bioluminescence.

Just as there are many kinds of paper, wood, waxes and other materials you could burn in air, there are many different types of luciferin in the sea, some restricted to particular types of organisms, others found in several different groups. While most bioluminescent organisms make their own luciferin, others, including some fish and squid, get their luciferin from bacteria growing in their light organs. Almost all these luciferins produce blue light, quite similar to the part of sunlight that penetrates the sea most successfully; and the eyes of most deep-sea animals are, naturally, particularly sensitive to these wavelengths. But some animals produce green or yellow light, while a very small number produce red light.

Searchlights

A few deep-sea animals use light in a very 'sneaky' fashion. Most bioluminescence is blue, so most deep-sea animals naturally have eyes that are particularly sensitive to blue light. At the same time, many of their bodies are either red or black because neither colour reflects blue light very well, so they are difficult to see. But three genera of small, black, deep-living fish, *Malacosteus*, *Pachystomias* and *Aristostomias*, have evolved a cunning solution. These fish feed on bright red shrimps, sometimes almost as big as themselves, which has earned them the name 'widemouths'. They have two distinct light organs, one of them emitting conventional blue light and probably acting as a recognition mark for other members of the same species. The other light organ, beneath the eye and shining forwards, produces deep red light, which must be totally invisible to the vast majority of the widemouths' neighbours, including the red shrimps, but will be reflected very efficiently by their red bodies. *Malacosteus* and its relatives have evolved eyes that are also particularly sensitive to this red light. So as the shrimps that these fish feed on swim around in what seems to them to be absolute darkness, they must be totally oblivious of the fact that their hunters are looking for them with the equivalent of an infra-red searchlight.

LEFT *Aristostomias scintillans,* one of the bathypelagic wide-mouths or rat-trap fish with its enormous jaws.

OPPOSITE **Close up of the head of** *Malacosteus* **showing its two distinctive light organs beneath the eye. The small one emits 'normal' blue light, while the larger one produces deep-red bioluminescence.**

Manna from heaven

With rare exceptions, all the food, or fuel, that drives the biology of the deep ocean originates in the very thin, near-surface layer where the sunlight allows the plants of the phytoplankton to live and grow. How does this fuel get to the animal communities in the deeper layers 5–10 km (3–6 miles) beneath the surface? The obvious short answer is that it sinks. Indeed it does, but in what form does it sink and how fast? The answers to these questions are by no means simple.

The potential food for the deep-sea animals, the plants and animals of the surface waters, are mostly very small. There are clearly many more tiny phytoplankton cells than herbivores feeding on them, and many more little herbivores than the somewhat bigger predators that feed on the herbivores. At each step in the food chain there is a marked decrease in numbers so that the very largest animals, including the top predators, are much less abundant than the tiny ones. When these large animals in the surface layers die, their carcasses sink rapidly and some of them, at least, fall all the way to the bottom to provide food for the sea-floor inhabitants. There are a number of deep-sea scavengers that have evolved the ability to exploit these large lumps of food when they finally reach the bottom.

A long journey

By far the largest source of food, however, sinks from the surface layers in the form of very small particles simply because phytoplankton cells, the tiny herbivores and the small predators are all much more abundant than large animals such as tuna, sharks and whales. Small particles sink much more slowly than large ones, so it was assumed until relatively recently that the vast bulk of the food leaving the surface layers and sinking into the deep ocean did so as a fine drizzle of tiny pieces sinking very slowly. From laboratory experiments and from calculations based on well-known physical principles, scientists concluded that many of these little particles would sink through the ocean at rates of no more than a few metres each day. This would mean that each tiny piece of food might take months or even years to travel the 5 km (3 miles) or more to the deep-sea floor.

Such a long journey time would have a number of very important consequences. First, over this sort of time-scale there seemed a very good chance that any particular food particle would never actually reach the deepest parts of the ocean, either being eaten by mid-water animals on the way or simply gradually decomposing into simpler and simpler chemicals until it eventually disappeared into solution in the water column. Second, in most parts of the ocean the horizontal water currents, including the relatively slow currents of the deep circulation system, would be very much faster than the vertical sinking rate of the small particles. So by the time it reached the bottom, a food particle might be transported tens or even hundreds of kilometres away from the place where it left the surface. This would mean that the biologically richest surface regions might not overlie the richest deep-sea regions and that the patterns of rich and poor animal communities on the deep-sea floor would have only the vaguest relationship with the surface patterns. Finally, any peaks and troughs in the supply of food particles as a result of seasonality in the surface layers would have disappeared by the time the slowly sinking particles reached the deeper parts of the ocean. We would, then, expect no seasonality in the deep ocean, where the animals would surely breed continuously, unlike many of their shallow-living relatives.

As knowledge of deep-sea biology improved during the middle decades of the 20th century, all these ideas seemed to be wrong. In many areas, the animal populations on the deep-sea floor were found to be richer than could be supported by a slow rain of tiny particles. The correspondence between the amount of life on the bottom and in the surface layers

above was not as poor as might be expected if all the food particles supplying it sank very slowly. Finally, and much more significantly, while many deep-sea animals were, indeed, found to be continuous breeders, a significant minority turned out to breed seasonally. How and why should they do this in a totally dark and constant environment where the food supply arrives at a constant rate?

Seasonal variations

The answer came as recently as the 1980s, when scientists discovered that in many areas of the ocean the amount of material sinking to the bottom varies dramatically with the time of year. For reasons that are still not understood fully, some of the algal cells in the

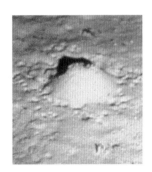
May 1

June 15

June 22

June 29

July 15

August 11

LEFT **Phytodetritus arriving on the deep-sea floor. A series of photographs of the floor of the northeast Atlantic at a depth of 4000 m (13,120 ft) taken between May 1 and August 11. There is almost no change throughout May and June. Then suddenly, between late June and mid July, a thick layer of phytodetritus sinks onto the seafloor, almost obliterating the mound which is more than 10 cm (4 in) high.**

RIGHT **Phytodetritus provides a welcome food 'bonanza' for many deep-sea animals. Here a holothurian echinoderm or sea cucumber, *Benthogone rosea*, plows through the phytodetritus eating its fill and leaving behind a faecal cast as it does so.**

surface waters clump together to form flocculent lumps sometimes up to several centimetres (an inch or so) across. These lumps sink rapidly through the water column at well over 100 m (330 ft)/day and so can reach the abyssal sea floor in a matter of two or three weeks compared with the many months that an individual algal cell would take.

This phenomenon probably occurs in most parts of the world's oceans, but its effects are most dramatic where the growth of the phytoplankton is particularly seasonal. In these areas the spring bloom of the algal populations can outstrip the ability of the herbivores to graze them, leaving a large surplus of algal cells. When these cells clump together they sink through the water column like a marine snow storm, picking up other material on the way, including dead and dying mid-water animals, moulted skins of crustaceans, faecal pellets and so on.

When this material, called phytodetritus, first reaches the bottom it can accumulate over a few days into a uniform flocculent carpet of nutritious 'fluff', sometimes several centimetres (an inch or so) deep. As it is wafted around by the near-bottom water currents, its distribution becomes more and more patchy, with some areas, particularly depressions in the sea bed, receiving large amounts of the material, while others receive very little or none at all. Apart from providing a mechanism by which changes in the availability of food in the surface layers can be reflected rapidly in the deep ocean, giving at least some of the deep-living animals a reason for breeding seasonally, the phytodetritus also seems to contribute to the small-scale patchiness of the deep-sea floor. This is crucially important, because it means that, as far as the very small sea-floor animals are concerned, the deep-sea bed is a much more complicated environment than anyone imagined just a few years ago. This, in turn, may at least partly explain one of the other great surprises of recent years, the remarkable species-richness of the deep sea-floor communities.

OPPOSITE **The richness of life around coral reefs is legendary. But the deep-sea floor may be even more species-rich.**

Biodiversity in the deep ocean

Biodiversity was one of the most powerful environmental buzz words of the late 20th century, with major international conferences devoted to discussing it and dozens of governments signing up to agreements to try to maintain it. But what does it mean, and what has it got to do with the deep ocean? Broadly, biodiversity refers to the variety of living forms with which we share the planet Earth, ranging from the genetic diversity within individual species of animals or plants to ecosystem diversity, such as deserts and wetlands, forests and savannahs. Within this range, a relatively simple feature – the number of species found in any particular area or environment – is often used to measure and compare biodiversity. In this sense, classical high-biodiversity environments include tropical rain forests on land, with their amazing variety of plants and animals, and coral reefs in the seas, with their equally impressive communities of invertebrates and fish. With the exception of coral reefs and one or two other marine habitats, the oceans have been considered to be much less biodiverse than the land.

Life on Earth almost certainly originated in the sea and was more or less restricted to the oceans for the first 3 billion years of evolution, so it is perhaps not surprising that in terms of the sheer variety of animal form, the marine environment is much richer than the land. The seas contain representatives of no less than 28 phyla – the major groupings of animals, such as sponges, crustaceans, molluscs and so on – whereas terrestrial environments have only 11 phyla. The situation is very different, however, at the species level. The number of marine plant species (4000 or so) is very small compared with the estimated 250,000 species of terrestrial green plants. Similarly, of the 1.7 million-or-so known animal species, only about one-tenth are from the oceans, with most of these living on or close to the bottom in relatively shallow water.

No barriers

The mid-water realm in the sea, including the deep ocean, seems to be particularly species-poor compared with the land or fresh water. For example, the copepods, the dominant animal group in the plankton throughout the oceans, are represented by fewer than 2000 species, whereas among the insects, the copepods' nearest terrestrial equivalents, there are several hundreds of thousands of species in just one order – the beetles. Similarly, despite the enormous size of the oceans compared with fresh-water systems, only just over half of the estimated 25,000 fish species on Earth are marine; most of these live in shallow, warm waters, with the vast deep-ocean water column harbouring only 1000 or so species.

The reason for this relatively low species-richness in the open sea seems to be that the oceans have fewer major physical or ecological barriers than terrestrial and fresh-water environments. As a result, many mid-water marine species are very widely distributed, with quite a few being found throughout the world's oceans. This means that there are far fewer opportunities for populations of animals to become separated and to evolve in different directions, possibly eventually giving rise to new species. At the same time, at least in mid-water, there are fewer specialist roles to be occupied by individual species than there are on land, so there is less encouragement for the evolution of lots of different species. For these reasons, while there are doubtless many more mid-water animals to be discovered, it seems likely that the total number of species in this environment will not exceed 10,000–20,000.

Species rich

In contrast, the sea-floor communities are much more diverse. Almost 200,000 sea-floor species have already been described and there are almost certainly many thousands more species to be found on the sea floor. A few thousands of these will be found in relatively shallow waters, particularly in areas of the world that have not been well studied. A few

BELOW **Deep-sea multiple corer being deployed from a research ship. This strange looking instrument has revolutionised deep-sea biology since it was developed in the 1970s. Unlike previous sediment samplers, it takes totally undisturbed samples of the bottom mud with the sediment-water interface intact.**

will be quite large, some perhaps very large, including giant squid and fish. The vast majority, however, will be tiny creatures, no more than a millimetre or two (0.1 in) long, living in the muddy sediments at the bottom of the deep ocean. We predict this because in the 130 years or so in which the deep-sea floor has been studied, only a tiny proportion of it has been sampled – something like 10 km^2 (4 mile^2) out of a total of more than 300 million km^2 (116 million mile^2). Only a few thousand species have been described from these samples, and estimates of the total number of species present vary enormously. Scientists all agree that many species remain to be discovered, with some of them claiming that there may be as many as ten million or more.

This is an amazing change of heart because, until the 1960s, most marine scientists thought that the deep-ocean floor, just like the mid-water realm, was particularly species poor. What has happened to change their minds? First, deep-sea scientists have been studying smaller and smaller mud-dwelling animals and finding more and more species. At the same time, they have found that the deep-sea floor is not the monotonous environment that it was once thought to be. Instead, at the time- and space-scales relevant to these tiny animals, it is amazingly variable, or patchy. Large animals moving over the bottom churn it up into mounds and depressions; water currents move the sediment surface around, scouring it from some areas and depositing it in others; carcasses of large animals from the overlying water column sink to the bottom, producing very localized food hot-spots; other potential

LEFT **A sediment sample taken by a multiple corer showing an undisturbed piece of the deep-sea floor with traces of phytodetritus still lying on the surface. Within the mud there will be hundreds of tiny animals belonging to many different species.**

food falls to the bottom like a nutritious snow, but then accumulates in thick patches separated by relatively barren areas; and from time to time slumps and slides of sediment like underwater avalanches sweep everything away so that the whole colonization process has to start all over again. These are very like the processes going on in tropical rain forests and in coral reefs, which encourage the evolution of the species-rich communities typical of these environments. If they act in the same way on the deep-sea floor, this vast environment, once thought of as a lifeless void, may turn out to be the most biodiverse region on Earth.

Life on the deep-sea floor

Larger animals

If you were able to take a stroll across the floor of the abyss almost anywhere in the world's oceans, and assuming you were provided with a strong searchlight, you would be confronted by a very strange scene. The first thing you would notice is that the creamy-brown mud of the bottom stretches away from you in all directions, almost dead flat for as far as you could see. But if you looked more closely at the bottom near where you were standing you would notice that, at a small scale, the bottom was not flat at all. Instead, there would be many mounds and depressions up to 30–40 cm (12–16 in) across and 5–10 cm (2–4 in) high. Some of the mounds would be smooth and rounded, others would be conical with a hole in the top like mini-volcanoes and sometimes with a circle of holes surrounding the base. There would be lots of other holes in the mud, sometimes occurring singly, sometimes in clusters, and ranging from a few millimetres to a couple of centimetres across. If you happened to be there just after a mass of phytodetritus had reached the bottom, you might see greenish masses of material where it had been swept by the water currents to gather in the depressions and around the mounds. Finally, the surface of the mud would be marked by a variety of strange grooves and little pock-marks meandering across the bottom like tiny tyre tracks.

All these features of the sediment surface are the work of the sea-floor animals. Many

of the mounds and holes are made by different types of worms living in the mud and either stretching their bodies over the surface to feed, or throwing out the contents of their burrows to form the mounds. Other holes are produced by molluscs, crustaceans and even starfish and brittlestars doing much the same as the worms, while irregular gashes in the mud may have been caused by fish or shrimps as they pounced on a tasty morsel on the bottom.

Activity

The 'tyre tracks' are produced by larger bottom-living animals as they wander across the sea floor in search of food. If you followed one of the tracks you might eventually find a deep-sea crustacean related to the crabs and lobsters of shallow waters. Or you might come across a snail or a starfish slowly ploughing its way over the mud. It is more likely, however, that at the end of the trail there would be a sea cucumber, a strange relative of the starfish, brittlestars and sea-urchins. All these groups are echinoderms,

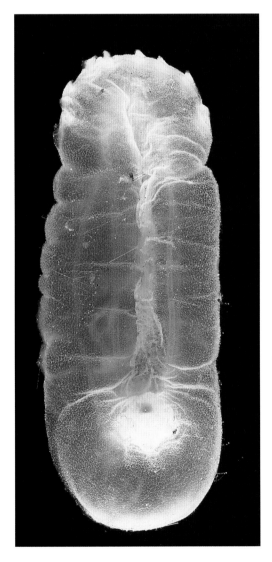

(meaning spiny-skinned). Whereas most of them are more or less circular, the sea cucumbers, or holothurians to give them their proper name, have more-or-less sausage-shaped bodies with definite front and back ends. Although sea cucumbers are well represented in shallow seas, they really come into their own in the deep ocean where they are one of the dominant sea-floor groups all over the world. They come in all sorts of shapes and sizes, ranging from tiny ones no more than a centimetre (0.4 in) long and living in burrows in the mud, through barrel-shaped versions with lots of stiff little legs, to massive football-sized giants, sometimes with long tails sticking up into the water. A few can even take off from the bottom and swim in the water column with remarkable undulations of their bodies. Almost all of them make a living by 'hoovering' the surface

ABOVE **Animals fresh from the deep-sea floor, including a large starfish, sea cucumbers, sea anemones and polychaete worms.**

LEFT **Sea cucumber, or holothurian** *Pencagone diaphana*, **from the deep-sea floor. Unlike most sea cucumbers, this one can swim using slow undulations of its muscular body.**

LEFT **This sea cucumber, _Oneirophanta mutabilis_, grows to about 15 cm (6 in) in length. It is one of the most abundant animals on the floor of the abyssal North Atlantic where it wanders around on its short peg-like 'legs' hoovering up small food particles from the sediment.**

BELOW **_Paelopatides grisea_ is a dorso-ventrally flattened sea cucumber which looks a bit like a pie crust. It moves across the sea bed with slow undulations of its 30 cm (12 in) long body.**

of the mud for edible scraps as they move across it and periodically dumping the unwanted contents of their guts as faecal casts. Because sea cucumbers are so universal, each species producing its own particular brand of trail and faecal cast, these are among the most common features on photographs of the deep-sea floor.

Because the deep-sea floor is poor in food resources, it cannot support large populations of big animals, particularly those species that use energy in moving around to find something to eat. So you would have to take ten or twenty steps, on average, to find a sea cucumber, a big crustacean, a starfish or other large mobile animal. Active free-swimming fish are even thinner on the ground. Many deep-sea animals adopt a 'sit and wait'

strategy, staying for long periods, sometimes permanently, in one place and hoping that food particles either fall on to them or are carried to them in the currents.

Patience

Many sit-and-wait animals project only a few centimetres (an inch) above the sediment surface. They include sponges, sea anemones, tube worms and even barnacles. There are some rather surprising members of the group; for example, one family of starfish, which are normally fairly active food hunters, sit still for long periods holding up their multi-branched arms to form a sort of basket to collect food particles drifting by. Many of the sedentary animals raise their feeding structures well above the bottom on long, thin stalks. So another strange feature of the vast abyssal plain is that, although there are no true plants, and nothing remotely resembling the size of a tree, here and there you would see odd plant-like objects with long thin stems, sometimes as much 1 m (3 ft) long, with a structures a bit like flowers perched on the top.

This sedentary life style is such a successful one in the deep sea that it has evolved quite separately several times. Some of the creatures you would see would be stalked sponges, others would be relatives of the sea anemones and corals called sea-pens, and yet others would belong to another group of strange sea-urchin and starfish relatives called crinoids or, because of their strong resemblance to plants, sea lilies. Clearly, all of them find standing still, and using very little energy to do so, helps solve one of the big problems of deep-sea life – finding food. However, it presents the animals with another, potentially equally serious, problem – how to find a mate and reproduce – because different individuals of these stalked species often seem to live tens of hundreds of metres apart.

ABOVE **Most starfish, whether in deep or shallow water, obtain their food from the sea bed itself. But some, like this one, _Freyella elegans,_ photographed at a depth of 4000 m (13,120 ft) in the North Atlantic, spread their arms into the water above the bottom and catch food particles carried in the water currents.**

LEFT **Most deep-sea sea anemones are rather small, no more than 2-3 cm (1 in) high. But this one, _Phelliactis hertwigi,_ grows to more than 10 cm (4 in) tall.**

In most of the species studied so far, the males apparently release their sperm into the water to fertilize the eggs within the female. Just how the male cells survive long enough to cross the large distances involved, and how they manage to find their way to a female with eggs, are two of the many deep-sea mysteries that still have not been solved.

Some animals of the deep-sea bed have no difficulty in covering long distances, either in search of a mate or a meal. In our stroll across the bottom we may come across evidence of this, an area of the sediment that appears to have been churned up like a ploughed field. Let us now take a break from our stroll and see if we can find out what could have caused this disturbance.

ABOVE Sea-pen, *Kophobelemnon stelliferum*. Like its coral relatives, this is a colonial animal, made up of many individual 'polyps', each one a bit like a sea anemone.

LEFT Crinoid echinoderm, *Bathycrinus gracilis,* at 4000 m (13,120 ft) northeast Atlantic. Crinoids are also called sea lilies because of their plant-like appearance. They are rather like starfishes on long stalks and feed on tiny food particles drifting past them in the near-bottom water currents.

LEFT *Eurythenes gryllus,* a deep-sea scavenging amphipod. Most amphipods attracted to dead carrion on the deep-sea floor are no more than a centimetre or two long. But *Eurythenes* is a giant, growing to about 15 cm (6 in) long.

Vultures of the abyss

Sooner or later, all the animals in the water column above the deep-sea floor die. For the little ones, such as copepods, shrimps and small fish, the end is likely to be sudden, as they are snapped up by predators. But the larger mid-water animals, such as squid, sharks and even dolphins and whales, must quite often die from disease, old age or even a heart attack. Let us now follow one such unfortunate creature and see what happens to it.

Sinking carcass

With rare exceptions, immediately after an animal dies its carcass will start to sink. At first, particularly if it is a fish with a swim bladder or a whale with a lung full of air, it may sink rather slowly. But as the pressure increases and any gas-filled space is squeezed smaller and smaller the carcass will gather speed until it is plummeting towards the bottom at a metre (3 ft) a second or more. Even so, it may take many hours to reach the sea floor and it may be attacked, and have pieces bitten off it, during the downward journey. But sooner or later, what is left will thump on to the bottom and send up a cloud of mud, which will slowly drift away with the current.

Within minutes the carcass will have its first ravenous visitor, almost certainly a shrimp-like amphipod. Amphipods are found at all depths in the sea and even in fresh water. They are usually no more than a couple of centimetres (an inch) long, but on the deep-sea floor some species reach a length

55

of 10–15 cm (4–6 in) and have sharp and powerful jaws, which can slice flesh like a surgeon's scalpel. They can detect the odour of a carcass carried on the water currents, possibly from hundreds of metres away, and rapidly home in on it.

Consequently, within hours of its arrival on the bottom our carcass will be surrounded by a cloud of hundreds or even thousands of amphipods, which, left alone, would eventually strip it to a bare skeleton. They will not be left alone, however, because there are other scavengers lurking on the deep-sea floor just waiting for the arrival of a food parcel. Although there are probably no deep-sea fish that are exclusively scavengers, several of them, including rat-tails, are not fussy about what they eat and will certainly be attracted to a carcass, as will be some of the deep-water shrimps. The amphipods will

BELOW **Rat-tail fishes,** *Coryphaenoides armatus,* **attracted to a baited camera at a depth of about 5000 m (16,400 ft).**

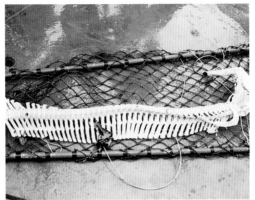

LEFT **A graphic demonstration of the work of deep-sea scavengers. This dolphin carcass was reduced to a bare skeleton after only five days on the bottom of the North Atlantic at a depth of almost 5000 m (16,400 ft).**

soon be joined by a number of other active swimmers and the sediment around the carcass will be disturbed rapidly by their feeding frenzy. When the amphipods, fish and shrimps have eaten their fill, or when there is nothing suitable left for them, the tiny pieces remaining will attract slow-moving animals, such as molluscs, starfish, brittlestars and sea cucumbers. This would have been the situation we came across when we found the disturbance marks on the sea floor. It will not stay that way for long, because the second stage scavengers will gradually obliterate the marks made by the swimmers and replace them with their own tracks and trails. Finally, bacteria will finish the demolition job, even breaking down bones, so that within weeks of the arrival of our carcass on the bottom there would be hardly any evidence that it had ever existed. When we came across the site we must have been there in its intermediate stage, when the active scavengers had already left, but the evidence of their activities was still there. Now let us carry on with our stroll and look at some of the less obvious animals of the deep-sea floor.

Smaller beasts

So far we have noticed only the animals big enough to be seen with the naked eye. These represent just a small fraction of the deep-sea-floor community because most animals of the deep-sea bed are tiny, a few millimetres (0.1 in) long at most, and they live hidden away in the mud at our feet. If we were to scoop up a handful or two of this abyssal mud from almost anywhere in the world's oceans, spread it thinly in a shallow dish and examine it carefully with a microscope, we would find a whole new living world and very probably several species still undescribed and unnamed. This is a world in which the animals are so small that, to them, sea water feels as thick and sticky as treacle does to us, and where to move 1 m (3 ft) could be the work of a lifetime. Many different animal groups are represented here, but just four or five groups dominate in terms of numbers.

First, at the lower end of the size range are the single-celled animals, the foraminiferans and their relatives. These very simple organisms basically consist of a blob of protoplasm, which feeds by engulfing bacteria

or other tiny pieces of organic matter.
By secreting a chalky shell, or by sticking together mud particles or pieces of other animals' shells into balls, tubes or multi-chambered little houses in which they live, these tiny animals produce a bewildering variety of strange shapes. They can be so numerous that they may outweigh all the other mud-dwelling, many-celled animals combined.

Another important group of single-celled animals are the Radiolaria. They have skeletons made of glass-like silica or strontium sulfate, often in radiating spicules – hence their name. Most of the 4000 or so

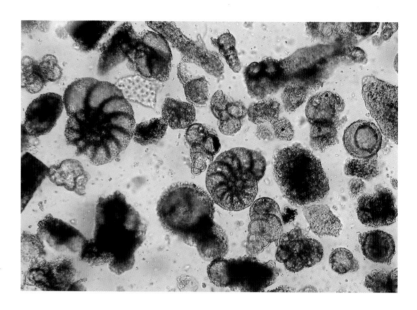

ABOVE *Acanthometra*, a radiolarian with skeletal spicules of strontium sulphate, and foraminiferans from the Indian Ocean.

TOP **Foraminiferan remains from the White Cliffs of Dover, UK. The cliffs are made up of unimaginable numbers of chalky shells of long dead marine animals.**

species are planktonic, but a few live on the bottom.

The second most abundant group, and among the smallest of the many-celled animals, are the nematodes or thread worms. Nematodes are found in all environments – marine, fresh-water and terrestrial – and may outnumber all other many-celled animals on Earth. In the deep sea they range in length from a few tenths of a millimetre to a centimetre or more and include bacterial grazers, specialist predators that can puncture the prey's cell membranes and suck out the internal juices, and active and deadly hunters. They have still not been studied extensively and many experts think that there may be tens of thousands of species yet to be described – so it is quite likely that our sample contains some of these unknown ones.

The crustaceans are the next in importance. They include amphipod shrimps, which are related to but much smaller than, the big

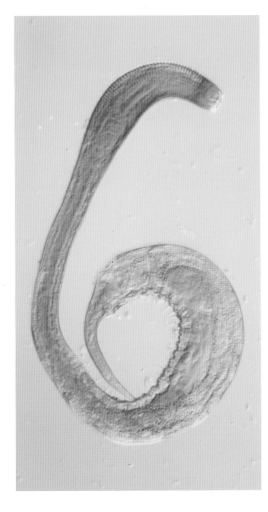

thin, matchstick-like bodies ideal for ploughing through the sediment; others with long spindly legs well-suited for walking across the soft surface of the mud. The most abundant of the mud-dwelling crustaceans, however, are the harpacticoid copepods, related to the mid-water copepods but specialized for living in the sediment and feeding on bacteria or organic detritus. Like the nematodes, many deep-sea harpacticoid species remain to be described.

LEFT **Marine nematode worm. Nematodes or thread worms are one of the dominant groups in deep-sea muds. There are certainly thousands, possibly tens of thousands, of unknown species still to be discovered.**

BELOW **Deep-sea isopods, *Eurycope gigantica*, related to the familiar woodlice or pill bugs.**

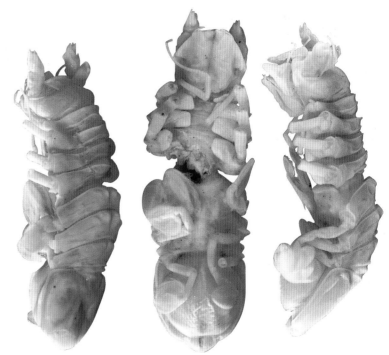

scavenging amphipods, and several other groups specialized in producing tunnels or in pushing their way through the sediment by simply moving aside mud particles or other animals. One of the most interesting of these groups is the isopods, to which the familiar terrestrial woodlice belong. Some deep-sea isopods resemble woodlice, but the group also includes an astonishing range of form: flattened creatures, which look like woodlice that have been stamped on; forms with long,

centimetres above the sediment surface which we might have noticed when we were looking at the larger animals. But in the sediments of the deep sea most of them are tiny, no more than a few millimetres (0.1 in) long. Even so, they include specialized carnivores, preying on the other tiny animals in the mud, as well

Next come the polychaete worms. These are true worms, belonging to the phylum Annelida, which includes the familiar earthworms. The polychaetes (meaning many bristles) are almost exclusively marine. They are found at all depths and include very mobile species, which crawl actively across the bottom or even swim, and sedentary species, which stay more or less in the same place, living in tubes or burrows and either foraging over the sediment surface for pieces of food or catching floating particles with feather-like tentacles. Some of them are quite big, and produce tubes sticking several

ABOVE **A newly discovered polychaete worm, *Sigambra* sp., from the floor of the abyssal Atlantic. Though some polychaetes reach 10-20 cm (4-8 in) most deep-sea species, like this one, are a few millimetres long.**

RIGHT **Hairy bivalves, *Limopsis pelagica* which live on or very near the sediment surface. The function of the hairs is unknown but may deter other animals from settling on the bivalves.**

as more general detritus-feeders such as the harpacticoids or hooverers, which simply swallow mud, digesting anything useful in it and ejecting the rest.

Only one other deep-sea, mud-dwelling group, the bivalve molluscs (relatives of the shallow-water clams and mussels with two shells, or valves, hinged together), approaches the polychaetes in abundance. We will come across giant bivalves when we look at hot-water vents on the floor of the Pacific, but most deep-sea bivalves are tiny, no more than a centimetre (0.4 in) across. Whereas many of their shallow-water relatives, such as clams and cockles, suck in water from above the sea floor and filter particles of food from it, this mode of feeding is rare among deep-sea bivalves. Instead, most of them live buried in the sediment and send out a feeding palp to gather pieces of food from the surface of the mud, while a few have become accomplished carnivorous hunters. Finally, and most surprising of all, there are specialist deep-sea, wood-boring bivalves. Their environment is just about as far from a source of any wood as you can get. These bivalves will obviously attack the hulls of sunken wooden ships, but they were around long before humans first sailed over the deep ocean. Unlikely as it seems, trees from coastal forests must fall into the sea and eventually sink into the abyss sufficiently often to make this strange lifestyle worthwhile.

ABOVE **Carnivorous deep-sea bivalves of the genus *Cuspidaria*. They burrow into the bottom mud and feed on small crustaceans.**

The very smallest

No account of deep-sea life would be complete without a reference to the smallest life forms of all, tiny organisms no more than a few thousandths of a millimetre long at most and until recently lumped together under the catch-all name of bacteria. These tiny cells occur everywhere on Earth where life is remotely possible, including in such unlikely environments as terrestrial hot springs with the pH of concentrated sulphuric acid and the coldest puddles in glaciers.

Their presence in the deep ocean has been known about for many decades, but knowledge of their variety, and importance, has grown enormously in recent years. They occur on and within all deep-sea animals, where they perform all sorts of crucial roles, ranging from producing bioluminescence to helping with digestion or even providing a source of food themselves. They also live independently of other organisms, being involved in a multitude of chemical processes in the water column and the sea-bed sediments, in and around hydrothermal vents, and associated with natural oil and gas seeps. Until the 1970s just a few rather similar types were recognized, but new techniques have revealed a bewildering variety in two quite distinct Kingdoms, the true Bacteria and a newly recognized one, the Archaea. Each of these Kingdoms is considered to be equivalent in rank to the third great division of life on Earth, the Eucarya, which includes all other living things, from protozoans and fungi to plants and animals. They have been found not only in the extreme environments of the hot springs, and the oil and gas seeps on the surface of the deep-sea floor, but have also been cultured from mud cores collected from hundreds of metres beneath the sea bed in sediments that must have been deposited millions of years ago. How they survive in this unbelievably hostile environment is still something of a puzzle, but the discovery of this deep biosphere is potentially enormously important. It suggests that tiny microbes may inhabit vast regions of the Earth's interior, way beyond the reach of any other life-forms. If so, despite their diminutive individual size the Bacteria and Archaea may in combination outweigh all the rest of the planet's living organisms put together. Unravelling the mysteries of this new-found, but ancient, world will be one of the great challenges to ocean scientists in the 21st century.

Fish of the deep-sea floor

There are about 1500 different fish species known from the floor of the deep sea and the waters immediately above it. By most standards they are rather ugly, and certainly

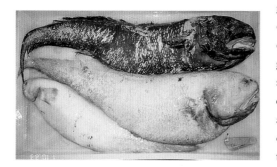

not very colourful, being mainly black, dark-grey, brown or various dirty-mud colours – and without even the light-coloured or silvery undersides typical of the camouflaging of fish from the sunlit upper levels. They fall broadly into two groups, which we can call the swimmers and the standers.

Swimmers

The swimmers are more or less neutrally buoyant, either because they have gas bladders to stop them sinking or because they have very light bones and flesh. This allows them to wander up and down in the water column in search of food without spending too much energy in this food-poor environment. They include members of many different fish groups but most of them are long and thin, often with particularly elongated tails. Many deep-sea fish are, therefore, rather eel-like in general appearance. The most abundant and species-rich deep-sea group, the macrourids, commonly called grenadiers, have large heads and taper more or less continuously to the tip of the tail so that they resemble large tadpoles. They are also called rat-tails – for

ABOVE *Caelorinchus caelorinchus*, a grenadier or rat-tail.

LEFT *Bassozetus levistomatus*, or cuskeel, a close relative of *Abyssobrotula*. These fish are unattractive even by deep-sea standards.

obvious reasons. Despite their strange appearance, they are distantly related to the cod family of shallow waters, and are represented by about 200 species worldwide at depths from about 250 m (820 ft) to the ocean trenches. They range in length as adults from about 20 cm (8 in) to 1.5 m (5 ft), and although some of them are specialized for scooping up mud from the sea floor and extracting tiny animals from it, most of them will eat more or less anything they can catch – dead or alive. Whenever they have been seen alive or filmed (you may have seen rat-tails swimming round the wreck of the ship in the film *Titanic*, for example) they seem

always to swim very slowly, so they probably catch their prey – crustaceans, molluscs, worms and other fish – by stealth rather than by speed and they are certainly attracted to dead carcasses lying on the bottom.

Although rat-tails are nowhere very common, and very few people have ever seen one, they may be among the most numerous animals of their size range on Earth because of the vast size of the deep-sea environment. One species, for instance, *Coryphaenoides armatus* (see page 56), seems to occur in all the major oceans of the world between about 3000 and 6000 m (9800–20,000 ft) depth and is probably represented by 20 billion individuals, more than three times as many as there are humans on the planet.

Standers

Unlike the swimmers, the standers tend to be heavier than water and sink if they stop swimming. This does not matter because they spend most of their time either resting on the bottom waiting for a suitable piece of food to arrive or making short excursions across the sea floor in search of food. They include relatives of the bottom-living skates and rays of shallow waters, and also bottom-dwelling sharks, although these are rarely more than about 1 m (3 ft) long.

Most of the bottom-dwelling bony fish are also fairly small and many of them, like the rat-tails, have massive heads and rather small bodies. The members of one family, the Liparidae, which are known for some reason

ABOVE **More deep-sea rat-tails, this time the long and short of** *Coryphaenoides* **species.**

RIGHT *Chimaera monstrosa*, and (bottom) *Hydrolagus* sp.

OPPOSITE **A long-nose chimaera,** *Rhinochimaera* **sp. The chimaeroid fishes, found in all the deep oceans of the world, are related to the sharks and have been around for more than 400 million years.**

BELOW *Paraliparis hystrix*, a sea snail.

as 'sea snails', are typical. There are many different species, but few grow to more than about 35 cm (14 in) long. One of them, *Careproctus amblystomopsis*, is one of the deepest-living fish on Earth, having been taken at a depth of 7230 m (23,700 ft) in the Kuril-Kamchatka Trench in the South-western Pacific. The deepest record for any fish goes to the undistinguished-looking *Abyssobrotula galatheae*, which grows to 15 cm (6 in) long, and is a member of another deep-sea family, the Ophidiidae (see page 62), which was dredged from the bottom of the Puerto Rico Trench at a depth of 8368 m (27,500 ft) in 1970.

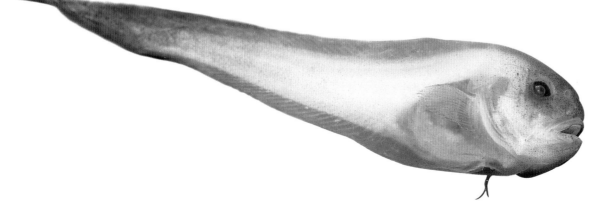

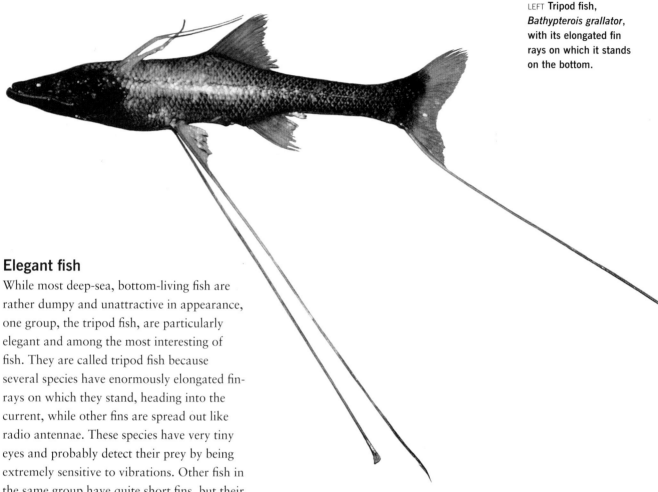

Elegant fish

While most deep-sea, bottom-living fish are rather dumpy and unattractive in appearance, one group, the tripod fish, are particularly elegant and among the most interesting of fish. They are called tripod fish because several species have enormously elongated fin-rays on which they stand, heading into the current, while other fins are spread out like radio antennae. These species have very tiny eyes and probably detect their prey by being extremely sensitive to vibrations. Other fish in the same group have quite short fins, but their eyes are enormous flat structures covering the whole of the upper surface of the head. Although these structures are certainly light-sensitive, they are not believed to be efficient as eyes, and these short-finned tripod fish probably rely on vibration sensitivity like their stilted cousins. All tripod fish have one other strange feature: each individual has both male and female sex organs. This phenomenon, called hermaphroditism, is by no means uncommon both in the sea and in other environments. But usually the testes and ovaries mature at different times so that the eggs of one individual are normally fertilized by the sperm from another, thus ensuring a mixing of genetic material. In the tripod fish, however, both sets of organs mature at the same time. This is possibly because the animals are so sparse on the bottom that they may not find a potential mate when they need one. If all else fails, one tripod fish can reproduce on its own by fertilizing its own eggs.

Monsters of the deep

RIGHT **Not a sea-serpent but a relatively small specimen of an oarfish proudly held by members of staff of The Natural History Museum, London, in the 1970s.**

RIGHT **Not a sea-serpent but a relatively small specimen of an oarfish proudly held by members of staff of The Natural History Museum, London, in the 1970s.**

Mankind has always been fascinated by tales of fearsome giant animals, and the mysterious deep ocean has been a fertile source of such stories. For hundreds of years, seafarers have brought back stories of strange and gigantic animals, usually some sort of sea serpent, that they have encountered on the high seas. Some of these accounts are undoubtedly based on misinterpretation of things such as tree trunks, whales, shallow-living sharks and other large fish swimming or floating at the surface, or even strange wave formations seen in conditions of poor visibility, for example at night or in bad weather. But others are certainly founded on real deep-sea giants, because they do exist.

The oarfish, or king of the herrings, *Regalecus glesne*, is an obvious candidate for sea-serpent sightings. With a flattened, snake-like body up to almost 10 m (33 ft) long, a bright-red dorsal fin and crest of long fin-rays on the top of its head, this strange and harmless fish is seen occasionally at the surface and may even be washed on to the shore. Because it is encountered so rarely, very little is known about the oarfish's life style. From its stomach contents it seems to feed mainly on planktonic animals. And because it is so large, it probably needs to live where its food is abundant, so it is likely that it does not live deeper than a few hundred metres.

Squid

Many myths of monsters must be based on sightings of squid, particularly the giant squid *Architeuthis dux*, which certainly do live deep in the ocean.

Squid are molluscs, related to mussels, oysters and snails. Along with the octopuses, they belong to a special group called the Cephalopoda, meaning 'head-footed', because in place of the flat 'foot' on which a snail crawls, cephalopods have a ring of long arms surrounding the mouth. The mouth in both octopuses and squid has a pair of powerful horny jaws a bit like a parrot's beak, but whereas octopuses live on the bottom, pulling themselves over rocks and in and out of crevices with the suckers on their eight arms, squid are mainly fast-swimming, mid-water animals whose ten arms, and particularly two extra long ones, the tentacles, are used to grab hold of food organisms and sometimes one another.

Squid bodies are generally cigar- or torpedo-shaped, often with a pair of fins at the tail end, away from the mouth and arms. Instead of an external shell like snails or mussels, squid bodies are stiffened by an internal skeletal rod, the pen, made of fairly pliable material more like cartilage than bone. They swim by jet-propulsion, sucking water into a space near the mouth, the mantle cavity, and shooting it out through a tube

LEFT *Histioteuthis pacifica*, one of the many species of small squids from the mesopelagic zone, in this case from about 500 m (1640 ft) deep near Hawaii. Almost all of these squids, like this one, have an impressive array of light organs.

RIGHT **Giant squid caught in a deep-sea trawl off the coast of Tasmania being unloaded in Melbourne. The animal was a 12 m (39 ft) long female.**

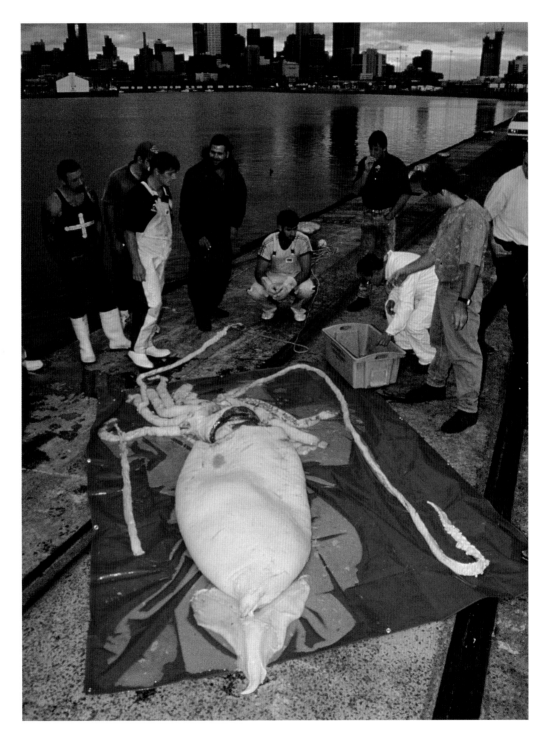

Octopuses

Not all octopuses live on the bottom. One deep-sea group, called the cirrate octopods, have soft, jelly-like bodies and spend much of their time in mid-water. Their name comes from the small outgrowths on their arms, called cirri, which are thought to be sense organs for detecting other animals in their pitch black deep–sea world.

But they also have other features which distinguish them from other octopods. First, they have a pair of fins which stick out from their rather globular bodies and presumably help them to swim. And several of them have the eight arms connected by thin webs, which make them look a bit like umbrellas when the arms are outstretched. In this position they can probably remain suspended in the water for long periods waiting for suitable food to chance by. Although they can be quite big, up to well over a metre across, cirrate octopods are far too slow moving and weak to catch fast swimming prey; so they probably feed mainly on small crustaceans and other planktonic animals that swim into their outstretched arms and become enmeshed in a sticky mucus produced near the mouth.

Vision doesn't seem to be very important to cirrate octopods; several species of the genus *Cirroteuthis* have greatly reduced eyes, while *Cirrothauma murrayi,* named for one of the scientists on the original *Challenger* expedition, is totally blind. Until recently, this seemed to fit in with the fact that no cirrate octopod was believed to produce bioluminescence. But another finned deep-sea octopus, *Stauroteuthis syrtensis,* has now been found to emit blue-green light from the suckers on its arms. So it seems likely that at least some *Cirrothauma* and *Cirroteuthis* species may do the same. And whether the octopods can see their light or not, it may be important in attracting prey animals, just like the lures on the 'rods' of angler fishes.

In contrast to the visually impaired cirrates, one deep-sea octopus relative has relatively enormous eyes, and a remarkable array of light organs all over its body that it seems to be able to turn on and off at will. Despite its ferocious sounding name, *Vampyroteuthis infernalis* (meaning 'the vampire squid from hell') grows to no more than about 20 cm (about 8 in) long and therefore probably feeds on rather small animals in the temperate and tropical lower mesopelagic zone (400-1000 m or 1320-3270 ft deep) where it lives. But although the water in this zone often has very low oxygen levels, observations from submersibles indicate that *Vampyroteuthis* is a good swimmer and can buzz around quite rapidly in search of food or to escape a predator. To do this, *Vampyroteuthis* is much more efficient than most other cephalopods at removing oxygen from the water. These strange features seemed to separate *Vampyroteuthis* from the other octopods with which it was lumped for many years after it was first described in 1903. Then it was found to have an extra pair of tiny 'arms' tucked into pockets outside the ring of main arms. But this did not mean that *Vampyroteuthis* was actually a ten-armed squid, for the extra arms turned out to be unique structures which are extended one at a time into long, fine filaments, rather like fishing lines, to detect prey which are then grabbed by the other arms. So little *Vampyroteuthis* seems to hover taxonomically somewhere between the eight-armed octopods and the ten-armed squids, yet another strange mystery from the depths of the ocean.

called a siphon by strong contractions of the mantle's muscular wall. By altering the direction of the outlet of the siphon they can move either tail first or arms and mouth first, as they must do when catching food.

There are several hundred different squid species in the oceans, ranging in size from tiddlers 2 cm (0.8 in) in length to the largest animals without backbones on Earth. Many of the shallow-water forms, including the ones we see on fishmongers' slabs, are tough-bodied, fierce, fast-moving hunters ranging up to half a metre (1.5 ft) or so in length. Most deep-sea squid, on the other hand, are at the small end of the size range and have rather soft and jelly-like bodies. They probably mostly wait for suitable food animals to bump into them rather than actively hunt them. There are a number of big deep-sea squid, however, which, although probably not as fast as their shallow-water cousins, make up for this in their awesome size. The biggest of all, up to at least 15 m (49 ft) long from the tip of the tentacles to the end of the tail and weighing a tonne or so, is *Architeuthis dux*, the giant squid.

Because it lives in the deep sea, and can presumably move out of the way of slow-moving nets, *Architeuthis* is hardly ever caught by scientists or fishermen. Most of what we know about the species has been gleaned either from its remains in the stomachs of sperm whales or from dead or dying specimens floating at the surface or washed up on to beaches. These strandings must have been happening for millions of years, ever since giant squids first appeared on Earth. But most of them have been, and

probably still are, unrecorded, either because no human being finds them, or because no-one realises what they are, particularly if they are badly damaged or decomposed.

The earliest recorded stranding appears to have been in Iceland in 1639. Not surprisingly

BELOW **Giant squid harpooned by the crew of the French corvette** *Alecton* **off the Canary Islands in 1861.**

the Icelanders at that time had no idea what it was, and it was more than 200 years before its true nature was recognised by a Danish zoologist in the mid-nineteenth century. Since then, there have been hundreds of strandings in many parts of the world, though for some unknown reason the greatest concentrations seem to be on the shores of Newfoundland in the Northwestern Atlantic, and around New Zealand. Many of the specimens, a few more or less complete but mostly rather badly rotted fragments, have found their way into museum collections where they have been studied and restudied by molluscan experts over many years. But although, as a result, the anatomy of *Architeuthis* is very well known, dead specimens tell us very little about how these amazing animals spend their lives. So almost everything written about the behaviour of giant squids is guesswork. For instance, we do not know exactly where *Architeuthis* lives, except that it is presumably at fairly deep levels. It has the largest eyes of any animal on Earth, up to about 30 cm (1 ft) across, the size of a dinner plate. This suggests that it hunts its food where the natural light level is very low, perhaps in the lower part of the mesopelagic zone from, say, 500–1000 m (1600–3300 ft). Of course, the eyes may be particularly good at seeing bioluminescence, in which case they could live in much deeper and darker waters, although unlike many other squid, *Architeuthis* does not produce any light itself.

So what does a monster squid feed on? Sperm whales have often been caught with scars on their skin clearly caused by giant squid suckers. These were taken as evidence of titanic battles between these leviathans of the abyss. If *Architeuthis* can tackle a 50-tonne (49-ton) sperm whale, it must surely be a ruthless and fearsome killer that can chase and capture any animal, tearing it to pieces with its lethal jaws. Probably not. For one thing, in any fight with a sperm whale an *Architeuthis* would be desperate to escape. Even the biggest squid is no match for a full-grown whale; it is simply not fast enough, powerful enough or even big enough to be a threat. So to the whale, the squid is probably simply a large and tasty potential meal with some irritating arms and suckers, which make it a trifle difficult to swallow. In fact, it seems likely that the giant squid is really a big lumbering softy, hovering in mid-water waiting for some unsuspecting potential food, maybe another squid or a biggish fish, to swim within reach of its tentacles. A big *Architeuthis* may need 50 kg (110 lb) or more of food each day, so it certainly qualifies as a major predator of the deep sea, but not as the diabolical killer and attacker of ships and sailors that mythology would suggest.

Deep sea 'Jaws'

If most stories of deep-sea serpents and monsters are based on these relatively harmless animals, is it possible that there are other fast-moving killers in the depth of the ocean – kinds of abyssal 'Jaws'? Reason suggests not. In the food-poor abyss a restless, voracious and speedy giant would probably not be able to find enough food to provide the energy needed to move around, let alone to grow and reproduce. This does not mean that there are no other deep-sea giants. For

OPPOSITE **A 3 m (11 ft) long Greenland shark,** *Somniosus microcephalus,* **beneath the Arctic ice.**

ABOVE **Megamouth,**
Megachasma pelagios.

example, the biggest known deep-sea fish is the Greenland shark, *Somniosus microcephalus*, which grows to at least 7 m (23 ft) in length and has been photographed at a depth of 2200 m (7200 ft). But *Somniosus* is not a Jaws. It is not an exclusively deep-sea species but also occurs close to the surface, where, because of its sluggish habits, mainly being attracted to offal thrown overboard from fishing vessels, it is also known as the sleeper shark.

It would be a mistake to assume that we already know all the deep-sea giants. As recently as 1976, American scientists working in the Pacific hauled aboard a shark 4.5 m (15 ft) long, which was quite different from any other known shark family. It was named *Megachasma pelagios*, meaning swimming big-mouth, partly because it had swallowed a large cargo chute that had been lowered from the ship as a sea anchor. Although the sea was about 1500 m (4900 ft) deep where *Megachasma* was caught, it had been swimming at only about 150 m (490 ft) down – hardly deep sea. Nevertheless, the fact that such a large and relatively shallow-living fish could remain totally unknown until so recently suggests that the deep ocean still holds many surprises for us.

Hydrothermal vents

Chimneys and toxic chemicals

Deep-sea scientists received a great shock in the late 1970s when springs of hot water were discovered gushing up through the sea floor into the icy waters of the deep ocean. The first of these hydrothermal vents were found by American scientists in the Pacific Ocean on part of the mid-ocean ridge system called the East Pacific Rise, but many more have been discovered since in all the main oceans and almost always associated with the crest of the ridge system at depths between 2000 and 3000 m (6600–9800 ft). The existence of such leaks of hot water through the ridges had been predicted by geophysicists for many years, but the form they took, and particularly the remarkable animal communities associated with them, took everyone totally by surprise.

The water coming out of the vents was once ordinary sea water that percolated downwards many hundreds of metres through cracks and fissures left in the new sea floor as it solidified and spread away from the ridge. At a depth of a kilometre (0.6 miles) or more into the sea bed the water met hot rocks of the oceanic crust and was both heated and chemically changed. The hot water was now very buoyant and rose, again through cracks in the rocks, to emerge into the sea through vents near the ridge crest. The temperature of the water coming out of the vents ranges from a modest 10–30°C (50–86°F), still much warmer than the 2–3°C (36–37°F) water

FAR LEFT **Black smoker with vent crab and fish.**

LEFT **A white smoker, at 2500 m (8200 ft) deep, on the East Pacific Rise, about 800 km (500 miles) southwest of Acapulco, Mexico. The white plume of hot water is gushing from its 30 cm (12 in) wide chimney, while a vent crab scurries away.**

normally found at these depths, to a
staggering 350–400°C (662–752°F). Because
of the very high pressure, this superheated
water does not boil and turn to steam as it
would in air. Instead, it simply gushes out of
the vent and mixes with the surrounding
water, usually cooling more or less to the
background temperature within a few metres.
When the hot vent water comes into contact
with the cold sea water, some of the chemicals
it picked up from the deep rocks may be
precipitated out as tiny particles making the
plume of water look a bit like smoke
billowing out of a chimney. Depending on the
temperature, the rising plume of water may
be black or white, and the corresponding
types of vents are called black smokers or
white smokers. They often produce their own
chimneys, tubes of precipitated chemicals
standing up to 30 m (98 ft) above the
surrounding sea floor before they become
unstable and crash on to the bottom to start
the process all over again.

Rather like the smoke issuing from
chimneys on land, the water gushing out of
deep-sea vents is laden with a cocktail of
chemicals that would be deadly to most forms
of life. While vents and the processes
producing them are fascinating to physicists
and geologists, it is the specialized organisms
associated with them that make the vents
really remarkable. These organisms not only
withstand the toxic chemicals but thrive in
them. Many vents are surrounded by fantastic
communities in which life may be hundreds of
times more abundant than on the adjacent sea
floor. This is because the vent animals have
their own rich source of food, which is totally

OPPOSITE **Massive 'chimneys' produced by a vent on the East Pacific Rise.**

independent of the input from the photo-synthesizing plants in the overlying surface waters. The food is produced by special bacteria, which, like the plants, can make the complex chemicals the animals need. But instead of using the energy in sunlight to do this they get their energy from some of the chemicals in the vent water in a process called chemosynthesis. Some of these bacteria are free-living and are eaten by other vent-dwellers; they can be so abundant that they form thick mats around the vents and even over the animals living there. Others live in close association with the animals, sometimes even inside their bodies, in a strange partnership from which both partners benefit.

Moving on

Although vents were discovered so recently, it seems that they have been around for many millions of years. In fact many scientists believe that their strange combination of chemistry, temperature and pressure make them prime candidates for the place of origin of the very first life on Earth. Individual vents are not long-lived, probably lasting no more than a few tens of years before they stop gushing forth their hot chemical soup as suddenly as they started. When this happens the community of animals living around them is doomed. This means that all the animals that live only near vents, and there are quite a few of them, have to be able to dispatch their young ones to find and colonize new vents. This is probably one of the reasons why the communities around vents that are reasonably close to one another on the same section of the ridge system tend to be very similar, but

quite different from those on other sections, and particularly in different oceans. Let us now look at a couple of the most dramatic of these communities and their amazing animal components.

Giant worms and monster clams

Molluscs such as mussels, cockles and oysters are all called bivalves because their bodies are enclosed inside two shells hinged together. There are many hundreds of bivalve species in the deep sea but almost all of them are tiny, no more than a few millimetres (0.1 in) long. So imagine the surprise of scientists looking at photographs from the deep Eastern Pacific in the late 1970s, which showed the sea floor strewn with empty bivalve shells up to 25 cm (10 in) long. These turned out to be the remains of vent communities that had died when the vent stopped working, but when the scientists in submersibles saw living vent animals for the first time they were totally amazed. In the vicinity of the vents they found not just a few big bivalves but hundreds of them crammed into crevices and often several deep. Along with them were other remarkable animals whose soft bodies leave no trace once the vents stop flowing.

The bivalves were of two distinct types, one related to the shallow-water mussels, the other to clams. The mussel, later named *Bathymodiolus thermophilus*, which simply means heat-loving deep-sea mussel, sometimes occurs in enormous numbers, up to $300/m^2$. This prompted the scientists to give the vents names like Mussel Bed and Clambake. They grow to a length of about 20 cm (8 in) and, like their shallow-living relatives, have well-developed gills and a functional mouth and gut. Also like shallow-living mussels they spend their lives anchored to the sea bed with a beard of tiny threads, the byssus, although from time to time they can detach themselves, move to a new spot, and become re-attached. So, like mussels the world over, they probably filter water through the gills and pass the

with its gills. Unlike the mussel, it moves about quite actively, using a large, fleshy foot, which it can push out from between its shells and either anchor itself into a crevice or pull its body along to a new location. Whereas when *Calyptogena*'s shallow-living relatives move in this way they are searching for food, the vent clam is probably searching for the best conditions to grow its gill bacteria.

FAR LEFT **Giant vent mussel** *Bathymodiolus*, **in this case** *B. elongatus*, **from a hydrothermal vent in the central Pacific at a depth of 2800 m (9180 ft). These animals obtain most of their nutrition from chemosymbiotic bacteria contained in their gills. These bacteria oxidise hydrogen sulphide emanating from the vents.**

LEFT **Shells of the giant clam,** *Calyptogena magnifica*, **from the East Pacific Rise.**

collected food particles, in this case vent bacteria, to the mouth. But the gut in *Bathymodiolus* is rather small and it seems that they have another source of food from the very high numbers of bacteria that actually grow inside the mussels' gills and attached to their surfaces. The bacteria seem to thrive in this situation even though many of them are destined to be eaten by their host.

The giant clam, *Calyptogena magnifica*, which grows even bigger than the mussel, also has a rich bacterial population associated

The clam has lost all trace of a gut and seems to be totally dependent upon its bacterial garden for its food.

If the scientists were surprised to see these giant vent bivalves, they were totally overawed by one of the other animals at the Pacific vents – tube worms almost 2 m (6 ft) long and as thick as a man's wrist growing in clusters like strange extra-terrestrial flowers that might have been found by Star-Trek voyagers on a previously unexplored planet. Named *Riftia pachyptila*, meaning the thick-feathered vent worm and referring to the brilliant red plume of gills sticking out of the top of the worm's white tube, these animals were placed by scientists in a totally new phylum, the Vestimentifera, which now has several other members found in other vent fields and around cold hydrocarbon seeps. Like the vent clam, *Riftia* has no mouth or gut. But up to half of its weight is made up of special chemosynthetic bacteria living in enormous numbers in the tissues of its body. The worm's blood delivers all the chemicals the bacteria need, having collected them from the vent water across the surface of the gill plume. The plume is richly supplied with blood vessels, which give its red colour.

Along with these giants, lots of smaller, and previously unknown, vent animals are found, including other molluscs and worms, shrimps, sea anemones and even fish. They are all fascinating in their own way, but one of the worms is particularly interesting because of its tolerance of very hot water. The Pompeii worm, *Alvinella pompejana*, is found particularly at hot vents in the Pacific, around white smokers where vent water emerges at around 150°C (302°F) and black smokers with water at 350°C (662°F). The worms grow to a length of about 10 cm (4 in) and have a mouth and a group of gills at one end and a series of branched hair-like outgrowths down the length of the body. These outgrowths carry a mass of chemosynthetic bacteria, similar to those associated with the bivalves, which provide a significant part of the worm's food supply. Each worm produces a thin-walled tube in which it spends most of its time with the gills sticking out of the end.

LEFT **Giant worms,** *Riftia pachyptila*, **clustered around a Pacific vent, and with a deep-sea squat lobster,** *Munidopsis*, **crawling over them.**

Giant worms

Yet another amazing feature of the giant worms is the rate they grow. What little we know of the growth of animals in the food-poor, deep sea away from vents suggests very slow growth rates indeed, with many of them, including some of the small ones, taking many years, sometimes decades, to reach full size. In stark contrast, *Riftia* appears to be able to do this in a matter of a couple of years or so, rivalling almost any animal in tropical shallow waters, let alone the deep sea. But science is rarely simple. In 2000, scientists working with a related species from hydrocarbon seeps in the Gulf of Mexico and also growing to a length of about 2 m (6 ft), found that this species apparently takes around 200 years or more to reach full size, making it the longest-lived (and probably slowest-growing) marine invertebrate apart from colonial animals like corals.

LEFT **Pompeii worm,**
Alvinella pompejana,
with its tube.

The tubes of hundreds of worms living together form a snowball-like mass on the side of the chimney close to the opening so that hot fluid flows through the tubes and over the worms' bodies. It is an incredibly dangerous place to live, within a few centimetres (an inch) of water that is so hot that it would cook the worms within seconds, and with the ever present possibility that the chimney will suddenly collapse and destroy the whole worm colony. Even more incredibly, the animals seem happy to sit with their tails at 70°C (158°F) or more and with the head at around 20°C (68°F), a gradient of no less than 50°C (90°F) along the length of the body. No other creature on Earth could stand such a gradient for more than a few minutes, so *Alvinella pompejana* amply deserves its name.

Ice worms

The combination of high pressures and low temperatures in the deep ocean has lots of other interesting effects – including making gas 'iced lollies'. Natural oil and gas (methane) seeps through the sea floor in lots of places, including the Gulf of Mexico. Here, at a depth of 650 m (2100 ft) the methane combines with water to form a strange, solid, ice-like material, although the surrounding water temperature is around 10°C (50°F). Associated with this peculiar stuff, in 1997 scientists found dense populations of a polychaete worm, related to well-known worms from more normal habitats but previously undescribed and clearly with a totally novel lifestyle. Christened 'ice worms', these creatures are now being studied to find out how they live in an environment that would be toxic to most other animals. Their planktonic larvae probably spend about 20 days in mid-water before they hopefully find another gas 'iced lolly' on which to settle.

Heat-seeking shrimps

Along with the giant worms and bivalves, the Pacific vent communities also include several shrimp species. Although they are interesting because they are found nowhere else, these Pacific shrimps are not very numerous or conspicuous. In the Atlantic, however, relatives of the Pacific shrimps dominate the vent faunas. There are no giant worms here, no big bivalves and no Pompeii worms. Instead, there are thousands upon thousands of white shrimps, about 5 cm (2 in) long,

LEFT **Shrimps swarming in their thousands at an Atlantic hydrothermal vent, along with crabs (*Bythograea* sp.) which are also found only at vent sites.**

which mass around the vents like bees swarming round a honey pot. There are a number of different species, found only at the vents, each with its own special life style and living in more or less distinct zones around the outflowing hot water. Despite the absence of any sunlight around the vents, several of them have the more or less conventional stalked eyes that are typical of most shrimps. A least one species, *Rimicaris exoculata*, appears at first sight to be totally eye-less.

Rimicaris lives in large swarms very close

It seems that the shrimp can clean these bacteria away with particularly flexible limbs and pass them to its mouth to be eaten. In fact it may be that this is the most important source of food and that the shrimps jostle one another to stay in water at the best temperature for their bacteria to grow. In other words, they seem to be farming their food on their own bodies. It is possible that, despite the absence of conventional eyes, the shrimps can 'see' exactly where they need to be to do this.

BELOW **The vent shrimp,** *Rimicaris exoculata*, **showing the modified eyes on the top of its head.**

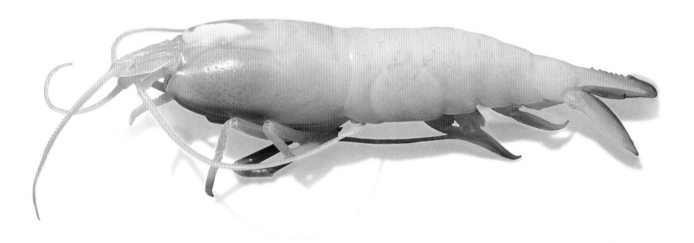

to the vent openings, each individual constantly jockeying for position with its neighbours. They feed on the vent bacteria living on the surfaces bathed by the outflowing water, but the shrimps are also covered in these bacteria, including the surfaces of the mouthparts or jaws and the lining of the gill chambers on either side of the animal's body above the bases of the legs.

Special vision

Each *Rimicaris* has a pair of strange organs on its back, connected to its brain by large nerves and lying just beneath the animal's transparent shell, or carapace. Despite being in quite the wrong place, these organs seem to be modified eyes. Although their structure would not allow them to see the sorts of images visible to normal shrimps, *Rimicaris* can certainly see something – but what? All hot objects emit radiation, which is why we can see metal glowing red- or white-hot. Most of this thermal radiation is not visible to us, but can *Rimicaris* 'see' and home in on the heat radiation coming from the black smokers where it lives? After lots of experiments it seems that the answer is probably not, because this radiation is simply too weak. Around the vents, however, there are several chemical reactions going on, which produce much stronger potentially visible radiation. It is still possible that totally 'blind' *Rimicaris* can 'see' the vents through eyes in the back of its head.

That *Rimicaris* eyes, along with those of other vent shrimps, definitely see something seemed to be confirmed recently by scientists who collected shrimp specimens from Atlantic vents that had been visited by manned submersibles only a month or so previously. When examined closely, the collected shrimps were found to have either pink eyes, with normal light-detecting structures in them, or white eyes in which these structures were broken down or totally destroyed. The scientists suggested that the white-eyed animals were ones that had been damaged by exposure to the submersible's flood-lights during the earlier visit. The pink-eyed forms had not been exposed to these lights during the previous visit, but would develop the abnormalities later as a result of the second visit. If this is true, then we humans may already have damaged at least some of the wonderful communities around the vents even before we know the answers to the many fascinating questions they pose.

How do vent animals find new vents?

One of the unsolved questions about hydrothermal-vent communities is how the animals find new vent sites to ensure the survival of their species? Individual vents are very short-lived in evolutionary terms, lasting for only a few tens of years. When the vent dies, so do the animals in the unique community around it. It is essential, then, that each vent species has very mobile juveniles, which can swim, or be carried by water currents to newly opened vents. The appearance of a new vent cannot be predicted, so these mobile young ones must be produced as often as possible just in case a new vent happens to become available. Most often, there will be no new vent. Some of the youngsters may find room to settle in vents that are already populated, possibly even the one where they were born. It seems, also, that very large carcasses, such as those of whales, may mimic some of the characteristics of vents and act a bit like stepping stones, enabling at least some vent species to bridge space- or time-gaps between vents. Nevertheless, many young vent animals must be doomed to wander the oceans until they are either eaten by a predator in mid-water or reach the stage when they must change into the vent-inhabiting adult form but fall to a vent-less sea bed where they perish. Either way, there must be a very high failure rate, so being a juvenile of a vent species must be extremely risky.

Even when a new vent is available and ready to be colonized, how on earth do the young ones find it? The answer is probably, largely by pure chance, although scientists do not actually know. A very slight clue was obtained in 1998 when scientists collected the tiny larval stages of a vent crab, *Bythograea thermodron*, from black smokers in the Pacific and kept the animals alive in a laboratory. They found that when the larvae were kept in cold water they were much more active and swam faster than when they were in warmer water, the reverse of what normally happens with most species. The scientists suggested that this behaviour would tend to keep the larvae moving in the cold water between vents, but would make them slow down and settle if they encountered warm vent water. Even the hottest vent water is cooled to very close to the normal deep-sea temperature within just a few metres of the vent opening, so the crab larvae must either be extremely sensitive to temperature or, more likely, their temperature-related swimming behaviour works only in the immediate vicinity of a vent. But how the larvae get this close to a vent, other than by chance, remains a mystery.

Man and the deep ocean

Mankind has been taking fish and shellfish out of the shallow waters over the continental shelves for hundreds of years. At the same time we have been throwing all sorts of rubbish into these inshore waters, including sewage, domestic and industrial waste and even unused military ammunition. We have caused a lot of damage. Fisheries have collapsed because of overfishing and the water off some beaches has become so polluted that no one dares to bathe in them. In contrast, until well into the 20th century, human impact on the deep ocean was much less than in shallow water. Apart from a few oceanic fisheries for fish such as tuna, and whaling in the past, almost no living resources were obtained from the open ocean. Likewise, except for waste thrown overboard from ships and, of course, the remains of sunken ships themselves, nothing was dumped into the deep ocean. But from about 1950 things began to change.

Natural resources
Fishing in deep sea

The limited biological productivity of the open ocean, and the problems of fishing in very deep water make it very unlikely that there will ever be any significant fisheries on the abyssal sea floor. Overfishing and resultant declines of shallow-water species, however, has already led to some commercial fishing on the fringes of the deep, at the top of the continental slope down to 1000 m (3300 ft) or more. Although several species are caught in these fisheries, by far the most important is the orange roughy, *Hoplostethus atlanticus*. Despite its scientific name, the orange roughy seems to live in all oceans at depths of around 800–1500 m (2600–4900 ft). It grows to about half a metre long and weighs up to 2 kg (4.4 lb) – and tastes very good. From time to time it occurs in very dense concentrations, particularly at breeding time and especially in the waters around New Zealand and Tasmania (Australia), where it has been fished intensively for the last 10–20 years. The effects of this fishery are very worrying. When the fishery started, very little was known of the orange roughy's biology or that it grows very slowly, possibly taking more than 70 years to

BELOW **Orange Roughy, *Hoplostethus atlanticus*, a handsome and widespread deep-living fish which is already suffering from overfishing in some places.**

reach full size. As a result, the fished populations have already been cut down to no more than one-fifth of their original size. In other words, they are overfished. This should be a warning not to begin commercial exploitation of other deep-sea species until much more is known about them.

Oil and gas

Drilling on land for oil and gas has been going on since the middle of the 19th century, but drilling beneath the sea dates back only to 1947 when the first truly offshore rig was installed in the Gulf of Mexico in just 7 m (23 ft) of water. Developments were rapid, and by the mid 1990s there were almost 6500

BELOW **A typical offshore oil and gas drilling platform.**

offshore oil and gas platforms, mostly on the continental shelves but a few in deeper water down to just over 1000 m (3300 ft). Although the technology is difficult and expensive, it is very likely that oil companies will want to drill at even greater depths in the next few years if the demand for oil, and its price, is high enough. Many environmentalists think this is wrong. They claim that the deep-sea activities harm the marine environment and, in any case, we should not be searching for more fossil fuel to burn and thus worsen global warming. Instead, we should be both reducing our energy needs and exploiting environmentally friendly sources of power such as wind, waves and, particularly, solar energy.

Metal-rich nodules and crusts

The world's industries are hungry for all sorts of materials, including a wide range of metals. Almost all of these are mined in land areas, but the deep ocean is a possible source for some of them. Large areas of the abyssal sea floor, particularly in the tropical Pacific, are littered with millions and millions of hard black 'pebbles' ranging in size from that of a hen's egg to a large grapefruit. Having formed over millions of years by chemical action, these manganese nodules contain not only manganese but many other metals, including nickel, copper and cobalt. These same metals occur in other areas of the ocean, not in nodules, but as more-or-less flat sheets or crusts, sometimes several metres across. Because of the technical difficulty of bringing them up from the bottom of the sea, none of these minerals has been exploited

LEFT **Manganese nodules on the sea floor of the deep mid-Atlantic. These nodules are about 5 cm across, the size of a small apple. In some parts of the Pacific Ocean the nodules are larger and more abundant.**

commercially yet. But many mining companies around the world are very interested in them and lots of ideas for collecting them have been proposed. If satisfying industry's needs for the metals from land sources becomes difficult, and therefore costly, enough in the next few decades, someone will undoubtedly try to exploit the nodules and crusts. To make such an operation commercially worthwhile it is

Wrecks

Although most wooden ships sunk in the deep sea will have long since disappeared, there must be many millions of tonnes of iron and steel shipwrecks on the bottom of the ocean. For example, between 1973 and 1995 alone, 29.3 million tonnes (28.8 million tons) of shipping were lost to accidental sinking worldwide, an average of 1.3 million tonnes each year. During wartime, of course, many more ships are sunk. Probably the most famous wreck of all time is that of the transatlantic passenger liner *Titanic,* which collided with an iceberg in the North Atlantic in 1912 and sank in 3800 m (12,500 ft) of water with a tragic loss of life. The wreckage on the sea bed was rediscovered in 1985 by a team led by the American scientist Dr Robert Ballard using submersibles, which obtained dramatic film of the remains of the ship, and of the sea-floor animals living in and around it; some of the film was used subsequently in a 'blockbuster' movie.

estimated that 1 km² (0.4 mile²) of sea bed would have to be cleared of nodules every day for 20 years. Millions of tonnes of unwanted mud brought up with the nodules would be released into the water column, so the environmental impact could be very serious indeed.

Deep ocean as a dump

During the second half of the 20th century the deep ocean was considered as a possible dump site for a range of mankind's nastier waste materials. Between about 1950 and 1980 many tonnes of low-level radioactive waste were dumped in the deep sea, particularly in the North Atlantic. This material, mainly contaminated gloves, paper towels, syringes and so on, was fairly harmless. Nevertheless, protests against the dumping eventually led to a total ban in 1993. By this time, proposals to dump much more dangerous medium- and high-level radioactive wastes in the deep sea, including spent fuels from nuclear power stations, had

ABOVE **Greenpeace activists attempting to prevent radioactive waste being dumped in the North Atlantic Ocean in 1981.**

also been given up in favour of other long-term solutions. Although the deep-sea option is unlikely to be considered again for at least 10–20 years, it may well be resurrected in the future if the alternatives prove to be too difficult or too dangerous. If so, it will be essential that any waste disposed of in the deep ocean is securely isolated, perhaps by burying it deep in the ocean sediments. Our knowledge of the oceans would also have to be good enough so that we can be absolutely sure that the waste will do no harm. Better still, we will have better solutions to the problems of radioactive waste and none of the Earth's environments, even one as far from mankind as the deep ocean, will have to be a home for it.

With the reduction in the use of shallow waters for the disposal of very bulky wastes such as sewage sludge and dredged material from clearing harbours and shipping channels, the deep sea has been considered as a possible disposal option. Compared with radioactive waste, sewage and dredge spoils are likely to be relatively harmless, and the deep ocean could probably swallow up large quantities with no detectable significant effects. Nevertheless, because of the bulky nature of these wastes, their disposal in the deep sea is likely to be a very expensive and technically difficult operation. They could not simply be tipped into the surface waters but would have to be deposited on to the sea floor in some way, either by pipe or carried to the bottom in enclosed containers. In any case, the general opposition to any marine waste disposal will probably ensure that it does not happen, at least for several decades.

Opposition to the oil and gas industry disposing of its redundant oil rigs and other offshore structures in the sea has also resulted in international regulations against it, so that, where at all possible, these structures must now be disposed of on shore.

Carbon dioxide

In general, then, the deep oceans seem safe from intentional waste disposal for the foreseeable future – except for one serious possibility. Since the industrial revolution, the burning of fossil fuels such as coal, oil and gas has caused big increases of greenhouse gases in the Earth's atmosphere, particularly of carbon dioxide (CO_2). These changes have almost certainly contributed to the increases in the atmospheric temperature, or global warming. The effect would have been much worse if the oceans had not already absorbed about half of the carbon dioxide produced by human activities and therefore reduced the amount left in the atmosphere. Unfortunately, no one knows whether the seas can go on taking up carbon dioxide in this way, so many environmentalists believe that we must reduce the amount we release into the air. Some people think that one way we could do this is to remove the carbon dioxide from smoke and other big sources and inject it deep into the ocean, either as a liquefied gas or in a frozen form rather like dry ice. At the temperatures and pressures found at the bottom of the ocean it is argued that the carbon dioxide will do no harm and will not get into the atmosphere to effect global warming, at least for several hundreds of years. Although this idea is attractive, no one knows for certain what would happen if large amounts of carbon dioxide were placed in the ocean in this way. Before it is tried a great deal more research must be carried out. Most people believe that a much better solution would be to produce less carbon dioxide in the first place. The problem is how to do this as the human population increases and the demand for energy, and therefore the burning of fossil fuels, grows every year. Alternatively, how do we convince people, particularly in the industrialized nations, to have simpler, less energy-hungry lifestyles?

And finally – who owns the ocean?

The possibility of human exploitation of the deep ocean throws up a question that has taxed mankind for centuries – who, if anyone, owns the seas? For more than 300 years, and until well into the 20th century, the generally held view was that each maritime nation had more or less total control over a narrow strip of sea bordering its coast, and that all of the rest was available for use by anyone. The width of these territorial waters, 3 nautical miles (5.5 km or 3.5 miles), was originally based on the range of 17th-century guns. But this same figure was kept long after guns could fire much greater distances, mostly because it suited those nations with big and powerful navies, such as Germany, the UK and the USA, who could do more or less what they wished in the rest of the ocean, the so-called High Seas.

By the middle of the 20th century, however, many less powerful nations with

LEFT **The world's exclusive economic zones (EEZ). Because of the projection used, the EEZs in high northern latitudes look larger than they should. But it is clear that much of the deep ocean, particularly in the central Pacific, is 'owned' by tiny island nations.**

coastlines were beginning to stake claims to much larger parts of their coastal waters. So an agency of the United Nations, UNCLOS (the United Nations Conference on the Law of the Sea) was established to draw up international laws about the use of the oceans. The UNCLOS has been sitting almost continuously for nearly 50 years, and is likely to go on meeting for many years to come. There are numerous extremely difficult problems and disagreements to resolve, including access to fisheries and control over pollution. But as far as use of the sea bed and its potential mineral resources are concerned, almost all nations now recognize coastal states' rights to Exclusive Economic Zones (or EEZs) stretching 200 miles (320 km) or more from their coastlines. As a result, almost half of the total area of the world's oceans falls within the EEZ of one nation or another and less than 60% is international High Seas. The EEZs are naturally mainly over the relatively shallow continental shelves and continental slopes bordering the major landmasses. But as the map opposite shows, large areas in the centres of the great oceans, including the abyssal Pacific, with its rich covering of manganese nodules, is covered by the EEZs surrounding dozens of tiny islands. Some of the Pacific islands belong to large nations such as France, the UK and the USA. But many of them belong to recently independent island states, such as Fiji, Kiribati, Tuvalu and Western Samoa, with very small human populations. Some of them are very vulnerable to rises in sea level as a result of global climate change. So if the industrialized nations of the world wish to exploit these deep-sea areas in the future, their governments may have to bargain with those of tiny nations with a quite different outlook. The next few decades could be a very interesting period in the history of man and the oceans.

Whether or not we succeed in solving these problems, and whether we use the deep oceans in some way to help us, one thing is certain: the oceans were here long before humans made their first appearance on Earth, and they will be here long after we have disappeared.

Glossary

Abyss Greek word meaning 'bottomless'. The deep part of the oceans, between about 3000 and 6000 m deep.

Abyssal plain The vast flat sea floor beneath the abyss. Almost always covered in fine mud.

Abyssopelagic The mid-water realm between the bottom of the bathypelagic zone at about 3000 m depth and the abyssal plain.

Amphipods Group of shrimp-like crustaceans ranging in length from a few millimetres to more than 10 cm.

Angler-fish Group of fish carrying on the top of the head a 'rod', developed from a modified fin-ray, which is dangled in front of the mouth to attract prey animals. In shallow-water anglers, including the monkfish or lotte, both sexes have rods and lures. In the deep-water species, belonging to a group called the Ceratioidea rods are found only on the females.

Arthropods The name, from a Greek word meaning 'jointed legs', given to a major animal group, or phylum, which includes insects, crustaceans and arachnids (spiders and their relatives).

Bacteria Tiny living organisms, neither clearly plants nor animals, which have no clear nucleus like the cells of most other creatures.

Bathypelagic The name, meaning 'deep sea', given to part of the third great mid-water depth zone, extending from the bottom of the 'twilight zone', or mesopelagic, at about 1000 m depth, to the top of the abyssopelagic at about 3000 m.

Benthic Greek word meaning 'depths of the sea' but applied specifically to the sea floor at all depths to distinguish it from the 'pelagic', meaning the water column above the bottom.

Benthos All the animals living within or on the sea floor.

Biodiversity 'Biological diversity', meaning the total range or variety of organisms found in the living world. It usually refers to the different types of animals or species found in a particular environment, but is also applied to within-species variety (genetic diversity) on the one hand, and community or habitat diversity on the other.

Bioluminescence 'Living light'. The dim, usually blue, light produced by many deep-sea animals. The light is sometimes produced by the animals themselves, and sometimes by bacteria living inside them. The process involves the oxidation of a substrate, luciferin, catalysed by luciferase.

Black smokers The name given to some very high temperature hydrothermal vents in which the hot liquid contains lots of brown or black particles making it look like smoke.

Carapace Part of the external skeleton or skin covering the head and back of arthropods including the crustaceans.

Catalysed Many chemical reactions are catalysed, that is made to work, by substances called catalysts. In bioluminescence the catalyst is a substance called luciferin.

Cell division The process by which most organisms grow, dividing one cell into two, two into four and so on, involving division and replication of the genetic material within a cell's nucleus.

Chemosynthesis Greek word meaning 'putting together chemically'. The process by which special bacteria build up complex substances using the energy from chemicals instead of using the sun's energy.

Coccolithophorids One of the most important groups within the oceanic phytoplankton. They are very small (a fraction of a millimetre across) single-celled plants, which can swim through the water using two whip-like flagella and are covered with tiny calcareous or chalky plates called coccoliths.

Comb-jellies or ctenophores Jelly-like planktonic animals, which propel themselves with rows of little hairs arranged like combs, and catch and eat other planktonic animals using small stinging cells similar to those of jelly-fishes.

Continental rise The bottom part of the continental slope, where the deep-sea floor begins to flatten out on to the abyssal plain.

Continental shelves The shallow parts of the seas adjoining the main land masses.

Continental slope Fairly steeply sloping part of the sea floor between the edge of the continental shelf, at about 200 m, and the continental rise at 3000–4000 m.

Copepods One of the most important groups of planktonic crustaceans in the oceans. Most are small, just a few millimetres long, and many feed on phytoplankton cells.

Crustaceans One of the divisions of the arthropod phylum. Marine crustaceans include the crabs, lobsters and shrimps, as well as many smaller creatures including the copepods. All arthropods are encased in a more or less rigid external skeleton, which has to be shed periodically to enable the animal to grow into a new and larger 'skin'.

Diatoms Planktonic single-celled plants in which the cell walls are impregnated with glass-like silica instead of the chalky or calcareous material found in other phytoplankton groups such as the coccolithophores.

Diel vertical migration The upward and downward movements made each day by many planktonic animals, mostly triggered by changing light intensity in the water column.

Dinoflagellates Phytoplankton group usually intermediate in size between the smaller coccolithophores and the larger diatoms.

East Pacific Rise Part of the mid-ocean ridge system famous as the location of the first hydrothermal vents discovered in the 1970s.

Echinoderms Phylum of invertebrates, which includes the sea urchins, starfishes, brittle stars, sea cucumbers and sea lilies.

Echo-sounding The process of determining the depth of the sea by sending out a sound signal from a ship and calculating the depth from the time taken for the sound to travel to the bottom and bounce back, and a knowledge of the speed sound travels through sea water.

Ecosystem All the characteristics of a particular living space, resulting from the interactions between the living organisms and the environment.

El Niño Name given to those years (usually at 4–10-year intervals) when atmospheric conditions over the subtropical Pacific Ocean reduce or even reverse the normal westward flow of the south Pacific gyre and the northward flow of the Peru Current that feeds it. The result is unusually warm waters off the coasts of Peru and Ecuador and often a failure of the fisheries.

Epipelagic Greek word meaning literally 'upon the sea'. Name given to the surface 100 m or so above the mesopelagic.

Faecal pellets Packages of waste material passed out of animal guts.

Fossil fuels Materials used by man as sources of energy, which were formed as a result of natural processes in the distant past. Oil, gas and coal are all fossil fuels.

Genetic diversity The variability within species, which produces the differences between individuals within a single species, rather than the broader-scale biodiversity resulting from differences between species.

Geophysics The study of the physics of the Earth, including processes such as sea floor spreading and hydrothermal vents.

Global warming The processes resulting in a rise in the average temperature of the atmosphere and the oceans.

Greenhouse gases Gases that contribute to global warming by allowing heat from the Sun to enter the atmosphere but not to leave it.

Gyres Circular movements of the oceans, particularly the enormous ones stirring the surface layers of the Atlantic and Pacific in clockwise directions in the Northern Hemisphere and anti-clockwise in the Southern Hemisphere.

Hadal Greek word meaning 'unseen'. Applied to the ocean trenches, with depths in excess of 6000 m.

Hatchet fish A group of small mesopelagic fish, so-called because they are extremely flattened from side to side and have silvered flanks that make them look like little axes.

Hermaphroditism Reproductive system in which the sexes are not separate, but both male and female organs are carried by the same individual.

Hydrothermal vents Springs of very hot, chemical-laden water gushing through the sea floor along the mid-ocean ridge system.

Industrial Revolution Name given to the period, roughly between 1750 and 1850, during which western human societies changed from a largely agriculture-based economy to a mainly industrial one.

Inorganic chemicals Chemicals derived from non-living sources, mostly rocks and minerals.

Invertebrates All animals without backbones or vertebrae. In the sea they range in complexity from the sponges and cnidarians (sea anemones, corals, jellyfishes and their relatives), through the worms and crustaceans, to the echinoderms and molluscs.

Lantern fish Group of mostly small mid-water fishes (myctophids), which get their common name from their silvery sides and many light organs.

Luciferase The chemical that catalyses the breakdown of luciferin to produce bioluminescence.

Luciferin The substance broken down under the action of luciferase to produce bioluminescence.

Manganese nodules Pebble-like objects, up to the size of a large grapefruit, found in large numbers in many areas of the deep-sea floor. They have been produced by the slow precipitation of chemicals over many hundreds of thousands of years. They consist largely of manganese oxide, but also contain many other metals of interest to industry.

Mesopelagic The middle zone of the deep-water column, extending from the bottom of the epipelagic zone to about 1000 m depth. Also known as the twilight zone.

Mid-ocean ridge system A more-or-less continuous, 45,000-km-long, submarine mountain range running through the world's oceans. The ridge is the site of production of new sea floor and of hydrothermal vents.

Organic chemicals Chemicals forming parts of living organisms or produced by them. A term also used generally for compounds containing carbon.

Oxidation Chemical process in which oxygen is combined with another substance.

Parasite Any organism that lives on or in another creature and obtains its food from it. Most parasites cause the host some degree of harm, often eventually killing it.

Peru Current Northward-flowing current off the west coast of South America and forming part of the anticlockwise gyre in the Southern Pacific, which in most years includes a strong westward subtropical flow driven by the South-east trade winds.

Photosynthesis Greek word meaning 'putting together with light'. The process in which green plants use the energy in sunlight to build complex organic substances, from carbon dioxide and water.

Phyla (singular phylum) Greek word meaning 'race' or 'tribe' given to the main groups of animals sharing the same general body plan. Examples of invertebrate phyla are the annelids (worms), arthropods (insects, crustaceans and their relatives), molluscs (snails, clams and squids) and echinoderms (starfish, brittle stars, sea cucumbers and sea urchins).

Phytoplankton The plant plankton (as opposed to the animal plankton or zooplankton). Mostly made up of microscopic coccolithophores, dinoflagellates and diatoms ranging from a few thousandths of a millimetre to a few millimetres across, but also including bacteria and other simple cells without a well-organized nucleus.

Radioactive waste Any waste emitting potentially harmful radiation. High-level radioactive waste includes the spent fuel cells from nuclear power stations. Low-level waste, dumped in the deep Atlantic until the 1980s, included protective clothing and swabs from radiological hospitals and laboratories.

Salps Mid-water relatives of the sea-squirts commonly found on rocky shores. Salps usually have barrel-shaped bodies up to a few centimetres long and open at both ends. Sometimes several individuals remain joined together to form chains up to half a metre or so long.

Silica Glass-like material making up the skeletal structures of many deep-sea animals, including the radiolarians and some sponges.

Siphonophores Mid-water animals, related to the jellyfish, but with many individuals joined together and co-operating in a single colony, rather like corals. Different individuals have specialized functions such as swimming, feeding or reproduction.

Species The basic unit into which scientists divide the living world. Basically, the members of the same species can interbreed with one another, but not with members of another species.

Thermohaline The name (from the Greek, meaning heat and salt) given to the system of massive but slow currents deep in the ocean. This deep system is driven by differences in the density (i.e. weight) of different water masses. These depend largely on the temperature and salt content of the water, cold and salty water being heavier than warmer or less salty water.

Trenches The deepest parts of the oceans at hadal depths, that is greater than 6000 m.

West Wind Drift The major surface current that flows more-or-less continuously around the Antarctic continent from west to east.

White smokers Relatively low-temperature hydrothermal vents in which the hot water plume carries grey or white particles.

Index *Italicised numbers refer to captions.*

Further information

Further reading

Deep Atlantic: life, death, and exploration in the abyss, R. Ellis. Alfred A. Knopf, New York, 1996.

Deep-Ocean Journeys: discovering new life at the bottom of the sea, Cindy LeeVan Dover. Addison-Wesley, 1996.

Deep-Sea Biology: a natural history of organisms at the deep-sea floor, J.D. Gage & P.A. Tyler. Cambridge University Press, 1991.

Deep-sea Demersal Fish and Fisheries, N.R. Merrett & R.L. Haedrich. Chapman and Hall, 1997.

Developments in Deep-Sea Biology, N.B. Marshall. Blandford, 1979.

The Ecology of Deep-Sea Hydrothermal Vents, Cindy Lee Van Dover. Princetown University Pres, 2000.

The effect of changing climate on marine life: past events and future predictions, A.J. Southward. & G.T. Boalch, pp 101-143 in *Man and the Marine Environment*, S. Fisher [Ed]. University Exeter Press, 1994.

Monsters of the Sea, R.Ellis. Robert Hale, London, 1994.

Oceanography, an illustrated guide. C.P. Summerhayes & S.A.Thorpe. A Manson Publishing Ltd., 1996.

The Search for the Giant Squid, R.Ellis. The Lyons Press, 1998.

Internet resources

CSIRO Marine Research
http://www.marine.csiro.au/
[research in the sustainable use of Australia's marine resources, the ocean's role in climate, and the effective conservation of the marine ecosystem]

Monterey Bay Aquarium Research Institute
http://www.mbari.org/
[centre for research and education in ocean science and technology]

USA and France TOPEX / POSEIDON satellite partnership
http://topex-www.jpl.nasa.gov/
[details of monitoring global ocean circulation, and discovering the tie between the oceans and atmosphere to help improve global climate predictions through satellite imagery]

Smithsonian Insitution
http://seawifs.gsfc.nasa.gov/ocean_planet.html
[virtual exhibition of the ocean planet]

Southampton Oceanography Centre
http://www.soc.soton.ac.uk/
[explores marine science, earth science and marine technology]

Woods Hole Oceanographic Institution
http://www.whoi.edu/
[all aspects of marine science are explored, with particular focus on education]

NB Web site addresses are subject to change.

Credits

Unless otherwise stated all images copyright The Natural History Museum.

Front cover Dr Emma Jones; **back cover and title page** SOC; **Front cover (inset) USA edn. only** Dr Gerhard Jarms; **p.5** NASA; **p.6** Mike Eaton; **p.7** (top) J F Bourillet, IFREMER DRO-GM Laboratoire Environnements Sedimentaires, France, (bottom) Mike Eaton; **p.9** Gary Hinks; **p.10 & 11** Mike Eaton (p.11 adapted from an Open University publication); **p.12** NASA; **p.15** (top) Institut Océanographique de Monaco, (bottom) Nigel Merrett; **p.16** (right) © Michael Thurston, SOC; **p.17** (middle) Nigel Merrett; **p.18** Nicola Mitchell; **p.19** Andrey Gebruk, P.P. Shirshov Institute of Oceanology, Russian Academy of Sciences; **p.20** (top) SOC, (bottom) Manson Publishing Ltd; **p.21** Mike Eaton; **p.23** NASA; **p.26 & 27** SOC; **p.28** Kevin Raskoff ©MBARI 2000; **p.29** (top and bottom) SOC; **p.30** Dr Gerhard Jarms; **p.31, 32, 33** (top & bottom), 35, 38 & 39 SOC; **p.40** P.J. Herring/ Y. Kito; **p.41** SOC; **p.42 & 43** © P.J. Herring; **p.45 & 46** SOC; **p.47** Linda Pitkin; **p.48** Tony Rice; **p.49 & 50** SOC; **p.51** (left) SOC, (right) Andy Nunn; **p.52** (top & bottom), 53 (top & bottom), 54 (top & bottom) & 55 SOC; **p.56 & 57** Dr Emma Jones; **p.62** (middle & bottom), **p.65** (bottom) & 66 Nigel Merrett; **p.67** Mirror Syndication International; **p.68** Dick Young; **p.69** David Paul; **p.73** Nick Caloyianis/NGS Image Collection; **p.74** © Tom Haight; **p.75** (left) Woods Hole Oceanographic Institution, (right) Richard A. Lutz; **p.76, 79 & 80** Woods Hole Oceanographic Institution; **p.81 & 82** Andrey Gebruk, P.P. Shirshov Institute of Oceanology, Russian Academy of Sciences; **p.86** UK Offshore Operators Association Limited; **p.87** SOC; **p.88** Greenpeace **p.90** Mike Eaton, adapted from David A. Ross and Therese A. Landry, Woods Hole Oceanographic Institution.

SOC = Southampton Oceanography Centre

MABARI = Monterey Bay Aquarium Research Institute

Acknowledgements – with thanks to Monty Piede, Cindy Lee Van Dover and all those who commented on the early draft of the manuscript.